Palgrave Studies in Digital Business & Enabling Technologies

Series Editors
Theo Lynn
Irish Institute of Digital Business
DCU Business School
Dublin, Ireland

John G. Mooney
Graziadio Business School
Pepperdine University
Malibu, CA, USA

This multi-disciplinary series will provide a comprehensive and coherent account of cloud computing, social media, mobile, big data, and other enabling technologies that are transforming how society operates and how people interact with each other. Each publication in the series will focus on a discrete but critical topic within business and computer science, covering existing research alongside cutting edge ideas. Volumes will be written by field experts on topics such as cloud migration, measuring the business value of the cloud, trust and data protection, fintech, and the Internet of Things. Each book has global reach and is relevant to faculty, researchers and students in digital business and computer science with an interest in the decisions and enabling technologies shaping society.

More information about this series at
http://www.palgrave.com/gp/series/16004

Theo Lynn • John G. Mooney
Jörg Domaschka • Keith A. Ellis
Editors

Managing Distributed Cloud Applications and Infrastructure

A Self-Optimising Approach

palgrave
macmillan

Editors
Theo Lynn
Irish Institute of Digital Business
DCU Business School
Dublin, Ireland

John G. Mooney
Graziadio Business School
Pepperdine University
Malibu, CA, USA

Jörg Domaschka
Institute of Information Resource
Management
Ulm University
Ulm, Germany

Keith A. Ellis
Intel Labs Europe
Dublin, Ireland

ISSN 2662-1282   ISSN 2662-1290 (electronic)
Palgrave Studies in Digital Business & Enabling Technologies
ISBN 978-3-030-39862-0   ISBN 978-3-030-39863-7 (eBook)
https://doi.org/10.1007/978-3-030-39863-7

This Palgrave Macmillan imprint is published by the registered company Springer Nature Switzerland AG.
The registered company address is: Gewerbestrasse 11, 6330 Cham, Switzerland

PREFACE

This is the third book in the series, "Palgrave Studies in Digital Business & Enabling Technologies", which aims to contribute to multi-disciplinary research on digital business and enabling technologies, such as cloud computing, social media, Big Data analytics, mobile technologies, and the Internet of Things, in Europe.

Previous volumes sought to consider and extend conventional thinking on disrupting finance and on cloud computing architectures to greater support heterogeneity, and specifically high-performance computing workloads. This third volume focuses more specifically on distributed compute environments that encompass resources, services, and applications from the cloud to the edge. The emergence of the Internet of Things (IoT) combined with greater heterogeneity, not only online in cloud computing architectures but across the cloud-to-edge continuum, is introducing new challenges for managing applications and infrastructure across this continuum. The scale and complexity are simply so complex that it is no longer realistic for IT teams to manually foresee the potential issues and manage the dynamism and dependencies across an increasing interdependent chain of service provision. This book explores these challenges and offers a solution for the intelligent and reliable management of physical infrastructure and the optimal placement of applications for the provision of services.

The content of the book is based on contributions from researchers on the RECAP project, a European Union project funded under Horizon 2020 (recap-project.eu). RECAP commenced in 2017 and brought together ten project partners from six countries across Europe to create a new way to optimally provision distributed heterogeneous resources to deliver services. The RECAP goal was to investigate and demonstrate an intelligent means to optimally place and scale dynamic applications and to manage the physical resources that underpin such applications, while targeting lower costs and better quality of service (QoS). RECAP is a model-based methodology, encompassing a reference architecture, models, and proof-of-concept implementations. This book is an output of this joint research.

The book is organised around key research contributions from RECAP.

Chapter 1 introduces the context driving the need for more reliable capacity provisioning of applications and infrastructure in distributed clouds. While IoT offers the potential of tremendous value for the private sector, public sector, and society as whole, it introduces complexity of several orders of magnitude greater in an already complex feature space. Chapter 1 introduces RECAP, an architectural innovation to support reliable capacity provisioning for distributed clouds. It presents a high-level conceptual overview of RECAP and some of the major design concepts informing its design, namely separation of concerns, model-centricism, modular design, and support for the use of machine learning and artificial intelligence for IT operations. The remainder of this book is loosely organised around the four functional building blocks of RECAP followed by a series of case studies to illustrate how RECAP can be implemented modularly.

Chapter 2 defines and discusses RECAP's Data Analytics and Machine Learning subsystem. This chapter describes the infrastructure for the acquisition and processing of data from applications and systems, and explains the methodology used to derive statistical and machine learning models from this data. These models are central to the operation of RECAP and are an input to all other subsystems, informing run-time planning, decision making, and optimisation support at both the infrastructure and the application levels.

Chapter 3 introduces RECAP's Application Optimisation subsystem. Optimising distributed cloud applications is a complex problem that requires understanding a myriad of factors. This chapter outlines the RECAP approach to application optimisation and presents its framework for joint modelling of applications, workloads, and the propagation of these workloads in applications and networks.

Chapter 4 introduces the RECAP Infrastructure Optimiser tasked with optimal application placement and infrastructure optimisation. This chapter details the methodology, models, and algorithmic approach taken to augment RECAP Application Optimiser output in producing a more holistic optimisation, cognisant of both application and infrastructure provider interests.

Chapter 5 focuses on Simulation and Planning in RECAP. The size and complexity of modern infrastructures make a real-time testing and experimentation difficult, time-consuming, and costly. The RECAP Simulation Framework offers cloud and communication service providers an alternative solution while retaining accuracy and verisimilitude. It comprises two simulation approaches, Discrete Event Simulation (DES) and Discrete Time Simulation (DTS), that provide enterprises with information about optimal virtual cache placements, resource handling and remediation of the system, optimal request servicing, and finally, optimal distribution of requests and resource adjustment. This information can inform better infrastructure capacity planning while taking in to account performance requirements and constraints such as cost and energy consumption.

Chapter 6 presents four case studies each illustrating an implementation of one or more RECAP subsystems. The first case study presents a case study on infrastructure optimisation for a 5G network use case. The second case study explores application optimisation for virtual content distribution networks on a large Tier 1 network operator. The third case study looks at how RECAP components can be embedded in an IoT platform to reduce costs and increase quality of service. The final case study presents how data analytics and simulation components, within RECAP, can be used by a small-to-medium-sized enterprise (SME) for cloud capacity planning.

Dublin, Ireland   Theo Lynn
Malibu, CA, USA   John G. Mooney
Ulm, Germany   Jörg Domaschka
Dublin, Ireland   Keith A. Ellis

ACKNOWLEDGEMENT

This book was funded by the European Union's Horizon 2020 Research and Innovation Programme through the RECAP project (https://recap-project.eu) under Grant Agreement Number 732667.

CONTENTS

Notes on Contributors

Malika Bendechache is a Postdoctoral Researcher at the Irish Institute of Digital Business at Dublin City University. She received her PhD in Computer Science at University College Dublin (UCD) in the area of parallel and distributed data mining. Bendechache was previously a researcher at the Insight Centre for Data Analytics at UCD. Her research interests span across distributed systems, Big Data analytics, and simulation of large-scale distributed cloud, fog, and edge computing environments and associated applications.

Paolo Casari is a Research Assistant Professor at the IMDEA Networks Institute, Madrid, Spain. His research interests include many aspects of networked communications, such as channel modelling, network protocol design, localisation, simulation, and experimental evaluations. He has (co)authored over 200 scientific publications, is a Senior Member of IEEE, and is an Associate Editor for the *IEEE Transactions on Mobile Computing* and for the *IEEE Transactions on Wireless Communications*.

Jörg Domaschka is a Senior Researcher and Group Manager at the Institute of Information Resource Management at Ulm University. He holds a Diploma in Computer Science from FAU, Erlangen Nuremberg, and a PhD in Computer Sscience from Ulm University. His research interests include distributed systems, fault-tolerance, middleware platforms, and NoSQL databases. Current focus of his work lies on middleware and run-time systems for geo-distributed infrastructure and applications.

Keith A. Ellis is a Senior Research Scientist and Manager of Intel Labs Europe, Ireland. His research focus is orchestration and control in Cyber Physical Systems. Ellis has led and been actively involved with national, international, and European part-funded research—REVISITE, COOPERATE, IMR, WISDOM, RealValue, and EL4L, targeting various domains—built environment, industrial, water management, smart grid. and agricultural. He is the holder of multiple patents and author of several journal articles, book chapters, and conference papers. He holds an MSc in Tech and Innovation Management and a BSc (Hons) in Tech.

Patricia Takako Endo is a Postdoctoral Research Fellow at Irish Institute of Digital Business, Dublin City University, Ireland, and a Professor at Universidade de Pernambuco, Brazil. Her research interests include cloud computing, fog computing, Internet of Things, system availability, and data analytics. Her articles have appeared in over 110 publications in the above research areas.

Antonio Fernández Anta is a Research Professor at IMDEA Networks Institute. Previously he was a Full Professor at the Universidad Rey Juan Carlos (URJC) and was on the Faculty of the Universidad Politécnica de Madrid (UPM). He spent sabbatical years at Bell Labs Murray Hill and MIT Media Lab. He has more than 25 years of research experience and more than 200 scientific publications. He was the Chair of the Steering Committee of DISC and has served in the TPC of numerous conferences and workshops. He is a Senior Member of ACM and IEEE.

Christos K. Filelis Papadopoulos received his Diploma in Engineering degree from the Electrical and Computer Engineering Department of the Democritus University of Thrace, Greece, in 2010 and his PhD in Numerical Analysis and High Performance Scientific Computations from the same university in 2014. His research interests include preconditioned iterative methods, multigrid and multilevel methods, and parallel computing.

Johan Forsman received an MS degree in Computer Science from Luleå University of Technology, Sweden. He is a Product Manager and Principal Solution Architect at Tieto Product Development Services. Forsman has over 20 years of experience in development of mobile telecommunication systems and is currently involved in business opportunities in the emerging telecoms landscape, introducing NFV, 5G, and IoT technologies. His domain of expertise includes mobile networks and specifically radio access networks and virtualisation technology.

Frank Fowley is a Senior Research Engineer in the Irish Institute for Digital Business (IIDB) and previously held the same position at the Irish Centre for Cloud Computing and Commerce (IC4) in Dublin City University (DCU). His main research revolves around cloud architecture and migration. Prior to joining DCU, Fowley held a number of senior positions in telecom and ICT companies in Ireland and abroad. He holds an MSc in Security and Forensic Computing and a BSc in Engineering.

Rafael García Leiva is a Research Engineer at the IMDEA Networks Institute, Madrid, Spain. Before this appointment, he was a Research Assistant at the University of Córdoba, R&D Coordinator at Andago Ingeniería, and a Principal at Entropy Computational Services. His research interests lie in the areas of Big Data and machine learning.

Konstantinos M. Giannoutakis is a Postdoctoral Research Fellow at the Information Technologies Institute of Centre for Research and Technology Hellas. His research interests include high-performance and scientific computing, parallel systems, grid/cloud computing, service-oriented architectures, and software engineering techniques. His articles have appeared in over 80 publications in the above research areas.

George A. Gravvanis is a Professor in the Department of Electrical and Computer Engineering of Democritus University of Thrace. His research interests include computational methods, mathematical modelling and applications, and parallel computations. He has published over 200 papers and is a member of the editorial board of international journals.

Frank Griesinger is a Researcher and Software Engineer at the Institute of Information Resource Management at Ulm University. He holds an MSc in Computer Science. The focus of his research interest is on the modelling, tracing, and self-adaptability of highly connected and dynamic applications as well as description languages and execution environments for cloud native applications.

Hector Humanes received his degree in Software Engineering and master's in Embedded and Distributed Systems Software from Universidad Politécnica de Madrid, Spain. Previously, he worked for the System and

Software Technology Group, a research group of the Universidad Politécnica of Madrid. Since 2018, he has been the Technical Leader of the Innovation Department in Sistemas Avanzados de Tecnología, S.A (SATEC), a Spanish ICT company.

Thang Le Duc is a Senior Researcher at Tieto Product Development Services with more than 10 years of R&D experience in both academia and industry. He received his PhD in Computer Engineering from Sungkyunkwan University (SKKU) and previously worked as a Postdoctoral Researcher at SKKU and Umeå University. Prior to that, he had held multiple academic positions and worked as a senior engineer in different industrial projects. His research interests include data analytics, system/workload modelling, cloud/edge computing, and SDN/NFV.

Mark Leznik is a Researcher and PhD Candidate at the Institute for Organisation and Management of Information systems at Ulm University. He holds an MSc in Computer Science from Ulm University, with the focus on computer vision, computer graphics, and machine learning. His current research interests include time series analysis, data synthesis, and anomaly detection using neural networks.

Radhika Loomba is a Research Scientist with Intel Labs Europe. She holds a PhD and BTech (Hons) degree in Computer Science and Engineering. Her PhD thesis focused on collaborative mobile sensing and mobile cloud computing technologies. Her current research focus is on orchestration, analytics, and optimisation for Cyber-Physical Systems from a mathematical modelling perspective, and her research interests include cloud computing, SDN, fog and edge computing, distributed collaborative systems, control theory, orchestration, planning, and scheduling.

Miguel Angel López-Peña holds a BS degree in Computer Science from Universidad Carlos III de Madrid, Spain, and a master's from the Spanish Ministerio de Educación (EQF level 7). He is currently pursuing a PhD in Science and Computer Technologies for Smart Cities at the Universidad Politécnica de Madrid. Since 2005, he has been an Innovation and Development Manager with the Sistemas Avanzados de Tecnología, S.A. (SATEC), a Spanish ICT company.

Theo Lynn is Full Professor of Digital Business at Dublin City University and is Director of the Irish Institute of Digital Business. He was formerly the Principal Investigator (PI) of the Irish Centre for Cloud Computing and Commerce, an Enterprise Ireland/IDA-funded Cloud Computing Technology Centre. Lynn specialises in the role of digital technologies in transforming business processes with a specific focus on cloud computing, social media, and data science.

John G. Mooney is Associate Professor of Information Systems and Technology Management and Academic Director of the Executive Doctorate in Business Administration at the Pepperdine Graziadio Business School. Mooney previously served as Executive Director of the Institute for Entertainment, Media and Culture from 2015 to 2018. He was named Fellow of the Association for Information Systems in December 2018. His current research interests include management of digital innovation (i.e. IT-enabled business innovation) and business executive responsibilities for managing digital platforms and information resources.

Linus Närvä is a Software Engineer at Tieto Sweden Support Services AB. His domain experience includes radio networks, radio base station software, and cloud computing platforms.

Manuel Noya is a Co-founder and CEO of Linknovate. He holds an MSc in Materials Science and Technology, a BSc in Chemical Engineering, and a BSc in Materials Engineering. He is an International Fellow at SRI International (Menlo Park, CA). His research interests include materials science, and software technologies in the area of text and data mining applied to business intelligence.

Per-Olov Östberg is a Research Scientist with a PhD in Computing Science from Umeå University and more than half a decade of postgraduate experience from both academic research and industry. He has held Researcher and Visiting Researcher positions at five universities: Umeå University, Uppsala University, and Karolinska Institutet in Sweden; Ulm University in Germany; and the Lawrence Berkeley National Laboratory (LBNL) at the University of California, Berkeley, in the USA. He specialises in distributed computing resource management and has worked in the Swedish government's strategic eScience research initiative eSSENCE, research and innovation projects funded by the EU under the FP7 and H2020 programmes, and projects funded by the Swedish national research council VR.

Minas Spanopoulos-Karalexidis is a Research Assistant at the Information Technologies Institute of Centre for Research and Technology Hellas. His research interests include high-performance scientific computing, simulation methods, sparse matrix technologies, iterative methods, parallel and distributed systems, and static timing analysis.

Sergej Svorobej is a Postdoctoral Researcher in the Irish Institute of Digital Business at Dublin City University. Svorobej's research focus is on complex systems, modelling and simulation with specific emphasis on cloud computing applications and infrastructure. Prior to working on the Horizon 2020 RECAP project, Svorobej was a Researcher at the Irish Centre for Cloud Computing and Commerce and on the FP7 CACTOS project. Previously, he held roles in SAP Ireland and SAP UK. He holds a PhD from Dublin City University and a BSc in Information Systems and Information Technology from Dublin Institute of Technology.

Dimitrios Tzovaras is the Director (and Senior Researcher Grade 'A') of the Information Technologies Institute. He received a Diploma in Electrical Engineering and a PhD in 2D and 3D Image Compression from the Aristotle University of Thessaloniki, Greece in 1992 and 1997, respectively. Prior to his current position, he was a Senior Researcher on the Information Processing Laboratory at the Electrical and Computer Engineering Department of the Aristotle University of Thessaloniki. His main research interests include network and visual analytics for network security, computer security, data fusion, biometric security, virtual reality, machine learning, and artificial intelligence.

Peter Willis manages the Software Based Networks team in BT Applied Research. He has been researching and developing Network Functions Virtualisation since 2011. He published the first carrier NFV testing results in June 2012 and is co-inventor of the term "NFV". Willis is currently leading BT's research to improve NFV and SDN technology and its management. Willis previously worked on the development of PBB-TE, BT's 21st Century Network Architecture, and BT's Internet service.

LIST OF FIGURES

LIST OF TABLES

Towards an Architecture for Reliable Capacity Provisioning for Distributed Clouds

Jörg Domaschka, Frank Griesinger, Mark Leznik,
Per-Olov Östberg, Keith A. Ellis, Paolo Casari,
Frank Fowley, and Theo Lynn

Abstract The complexity of computing along the cloud-to-edge continuum presents significant challenges to ICT operations and in particular reliable capacity planning and resource provisioning to meet unpredictable, fluctuating, and mobile demand. This chapter presents a high-level

J. Domaschka (✉) • F. Griesinger • M. Leznik
Institute of Information Resource Management, Ulm University, Ulm, Germany
e-mail: joerg.domaschka@uni-ulm.de; frank.griesinger@uni-ulm.de;
mark.leznik@uni-ulm.de

P.-O. Östberg
Umeå University, Umeå, Sweden
e-mail: p-o@cs.umu.se

K. A. Ellis
Intel Labs Europe, Dublin, Ireland
e-mail: keith.ellis5@mail.dcu.ie

1
T. Lynn et al. (eds.), *Managing Distributed Cloud Applications and Infrastructure*, Palgrave Studies in Digital Business & Enabling Technologies, https://doi.org/10.1007/978-3-030-39863-7_1

conceptual overview of RECAP—an architectural innovation to support reliable capacity provisioning for distributed clouds—and its operational modes and functional building blocks. In addition, the major design concepts informing its design—namely separation of concerns, model-centricism, modular design, and machine learning and artificial intelligence for IT operations—are also discussed.

Keywords Capacity provisioning • Distributed cloud computing • Edge computing • Infrastructure optimisation • Application optimisation

1.1   Introduction

The objective of this book is to introduce readers to RECAP, an architectural innovation in cloud, fog, and edge computing based on the concepts of separation of concerns, model-centricism, modular design, and machine learning and artificial intelligence (AI) for IT operations to support reliable capacity provisioning for distributed clouds. The remainder of this chapter provides a brief overview of computing across the cloud-to-edge (C2E) continuum and the challenges of distributing and managing applications across geo-distributed infrastructure. This chapter also introduces some of the major design concepts informing the RECAP architectural design and provides an overview of the RECAP architecture and components.

P. Casari
IMDEA Networks Institute, Madrid, Spain
e-mail: paolo.casari@imdea.org

F. Fowley
Irish Institute of Digital Business, Dublin City University, Dublin, Ireland
e-mail: frank.fowley@dcu.ie

T. Lynn
Irish Institute of Digital Business, DCU Business School, Dublin, Ireland
e-mail: theo.lynn@dcu.ie

1.2   FROM THE CLOUD TO THE EDGE AND BACK AGAIN

The convergence and increasing ubiquity of wireless internet access, cloud computing, Big Data analytics, social and mobile technologies presage the possibilities of billions of people and things connected through mobile devices and smart objects in the cloud. This phenomenon is heralded as the coming of the fourth industrial revolution, the networked society, the Internet of Things (IoT), indeed the Internet of Everything. Connecting but a fraction of the 1.4 trillion "things" worldwide today is predicted to create US$14.4 trillion and US$4.6 trillion in private and public sector value, respectively, through accelerated innovation and improved asset utilisation, employee productivity, supply chain, logistics, and customer experience (Cisco 2013a, b).

Today, while we are moving towards a society whose social structures and activities, to a greater or lesser extent, are organised around digital information networks that connect people, processes, things, data, and social networks, the reality is still some distance away (Lynn et al. 2018). The dawn, if not the day, of the Internet of Things is here. Haller et al. (2009) define IoT as:

> A world where physical objects are seamlessly integrated into the information network, and where the physical objects can become active participants in business processes. Services are available to interact with these "smart objects" over the Internet, query their state and any information associated with them, taking into account security and privacy issues. (Haller et al. 2009, p. 15)

This definition largely assumes that smart objects (end-devices), ranging from the simple to the complex in terms of compute, storage, and networking capabilities, will interact with each other and the cloud to provide and consume services and data, but not necessarily at all times. Furthermore, these smart end-devices, e.g. smart phones or transport sensors, may move to different geographic areas where, for economic, geographic, or technological reasons, they cannot always be connected, yet will be expected to carry on functioning regardless. IoT embodies many of the drivers that see an increased move from cloud-centric deployments to distributed application deployments in the cloud or on the edge infrastructure.

Within the traditional cloud computing paradigm, processing and storage typically take place within the boundaries of a cloud and its underlying infrastructure, and are often optimised for specific types of applications and workloads with predictable patterns. Neither the cloud nor the networks connecting these objects to the cloud were designed to cater for the flood of geographically dispersed, heterogeneous end points in the IoT and the volume, variety, and velocity of data that they generate.

Fog computing and edge computing are two relatively new paradigms of computing that have been proposed to address these challenges. Fog computing is a horizontal, physical, or virtual resource paradigm that resides between smart end-devices and traditional cloud data centres. It is designed to support vertically isolated, latency-sensitive applications by providing ubiquitous, scalable, layered, federated, and distributed computing, storage, and network connectivity (Iorga et al. 2018). In contrast, edge computing is local computing at the edge of the network layer encompassing the smart end-devices and their users (Iorga et al. 2018). If one imagines a cloud-to-edge (C2E) continuum, data processing and storage may be local to an end-device at the edge of a network, located in the cloud, or somewhere in between, in "the fog".

As discussed, while fog computing and edge computing offer solutions for delivering IoT to industry and the masses, they introduce new and significant challenges to cloud service providers, network operators and enterprises using this infrastructure. These environments face a high degree of dynamism as an immediate consequence of user behaviour. Overall, this setting creates a set of challenges regarding how to distribute and run applications in such unpredictable geo-distributed environments. Similar demands are seen at the network edge given the growth of relatively nascent services, e.g. Content Delivery Networks. Spreading infrastructure out over large geographic areas increases the complexity and cost of planning, managing, and operating that physical infrastructure. Firstly, it raises the question of how much infrastructure of what type to place where in the network—a decision that must be made in advance of any service being offered. Secondly, applications deployed over large geographically distributed areas require a detailed understanding of the technical requirements of each application and the impact on the application when communication between an application's components suffers due to increased latency and/or reduced bandwidth. Thirdly, for a service provider along the C2E continuum, the question arises about which (parts) of the various applications in a multi-tenant setting should be operated at the edge and which should not be. This is of critical importance due to the potentially limited compute resources available at each edge location. To add to the complexity, some of these questions must be answered in advance with incomplete data on user demand while others require near real-time decision making to meet unpredictable and fluctuating user demands.

Incorrect placement decisions may result in inflexible, unreliable, expensive networks and services. This is more likely as the decision space becomes so complex; it is no longer realistic for IT teams to cost-effectively foresee and manually manage all possible configurations, component interactions, and end-user operations on a detailed level. As such, mechanisms are needed for the automated and intelligent placement and scaling of dynamic applications and for the management of the physical resources that underpin such applications. RECAP—an architectural innovation in cloud and edge computing to support reliable capacity provisioning for distributed clouds—is posited as such a mechanism.

1.3   Design Principles

This section outlines some of the major design concepts informing the RECAP architectural design, namely separation of concerns, model-centricism, modular design, and machine learning and AI for IT operations.

1.3.1   Separation of Concerns

Separation of concerns is a concept that implements a "what-how" approach to cloud architectures separating application lifecycle management and resource management where the end user or enterprise customer focuses its efforts on what needs to be done and the cloud service provider or cloud carrier focuses on how it should be done (Lynn 2018). At its core, the end user or enterprise customer focuses on specifying the business functionality, constraints, quality of service (QoS), and quality of experience (QoE) (together KPIs) they require, with minimal interference with the underlying infrastructure (Papazoglou 2012). To support a separation of concerns, a detailed understanding of the KPIs but also the relationship between the performance of the applications and underlying infrastructure, and the achievement of these APIs is required.

In multi-tenant environments, for example clouds and networks, the separation of concerns is complicated because the actors will, most likely, belong to different organisations (including competitors), have very different KPIs, different load patterns, different network topologies, and more critically, different priorities. Any architecture for reliable capacity provisioning, whether from an application or infrastructure perspective, across the C2E continuum must have mechanisms to support separation of concerns in an agile way.

1.3.2   Model-Centricism

Due to the complexity, heterogeneity, and dynamic nature of (i) the business domains in which enterprises, cloud service providers, and cloud carriers operate; (ii) the application landscape (including legacy and next generation applications); and (iii) the infrastructure in and upon which these applications operate and are consumed, a flexible software architecture is required that can evolve in line with business, application, and infrastructure requirements. Model-centricism is a design principle that uses machine-readable, highly abstract models developed independently of the implementation technology and stored in standardised repositories (Kleppe et al. 2003). This provides a separation of concerns by design, and thus supporting greater flexibility when architecting and evolving enterprise-scale and hyperscale systems. Brown (2004, pp. 319–320) enumerates the advantages of using models including:

- Models help people understand and communicate complex ideas.
- Many different kinds of elements can be modelled depending on the context offering different views of the world.
- There is commonality at all levels of these models in both the problems being analysed, and in the proposed solutions.
- Applying the ideas of different kinds of models and transforming them between representations provide a well-defined style of development, enabling the identification and reuse of common approaches.
- Existing model-driven and model-centric conceptual frameworks exist to express models, model relationships, and model-to-model transformations.
- Tools and technologies can help to realise this approach, and make it practical and efficient to apply.

To meet the needs of infrastructure providers as well as application operators, an understanding is needed on how the impact of load and load changes on the application layer influences the application's resource demands at the infrastructure layer and further, how competing resource demands from multiple applications, and indeed multiple application providers, impact the infrastructure layer.

From a high-level perspective, users impose a certain load on the applications; that load will change over time. At the same time, users have performance requirements for a given application. For instance, a lack of responsiveness from a website may make them switch while otherwise they would have stayed. The operators of that application want to ensure that some level of performance is guaranteed in order to keep their customers. Hence, it is their task to adapt the performance of the application to the amount of workload imposed by the users. How and whether this can be done depends on the architecture and implementation of the application. For distributed applications (that constitute a huge portion of today's applications), horizontal scaling increases the computational capacity. This, in turn, reduces queuing and keeps latency constant despite increasing workload. Moreover, for applications composed of multiple different components, it is important to understand how load imposed at the customer-facing components ripples through the application graph and impacts the loads on each and every component. Finally, to understand how much performance a component running on a dedicated hardware unit (e.g. processor type, RAM type, and disk type) can deliver under a specific configuration (e.g. available RAM and available cores), a mapping needs to be available that translates load metrics on the application level such as arrival rate of requests of a specific type to load metrics on hardware such as CPU used, RAM used, disk usage, as well as the performance achieved from it. In multi-tenant environments such as virtualised cloud and cloud/edge systems, the mutual impact of multiple, concurrently running components from different owners on the same physical hardware is critical.

A model-centric approach for capacity provisioning for distributed clouds requires at least six models—(1) user models, (2) workload models, (3) application models, (4) infrastructure models, (5) load translation models, and (6) Quality-of-Service (QoS) models (Fig. 1.1).

User models describe the behaviour of users with respect to the usage of individual network-based services. That is, they capture different types of users and their usage patterns over time. What is more, they also describe their movement over geographical regions such that it becomes possible to understand which edge parts of the network will have dedicated

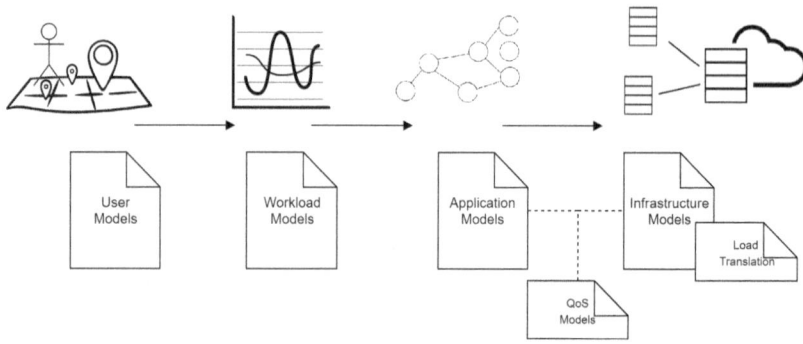

Fig. 1.1   Interdependencies between models

demands for specific services. This is of special interest to edge computing systems as user mobility impacts network load and application access patterns.

Workload models describe the workload issued on a system from users and external systems. While the user model captures the location and type of users, the workload model describes what actions these users execute and how this translates into interaction with which parts of an application.

Application models fulfil multiple purposes. First and foremost, they describe which components compose a distributed application and how these components are linked with each other (static application topology). This part of the application model also captures how to technically install the application in the infrastructure and how to update a running deployment. Deploying an application creates a run-time application topology that describes how many instances of each application component are currently available at which location and how they communicate with each other on a per-instance basis. The **(work)load transition models** as a sub-model of the application model describe for the application how incoming workload propagates through the applications' components and the impact this has on the outgoing links of the component.

As application models are not capable of determining whether or not a given application topology (or scaling factor) is capable of servicing a certain amount of load, as they neither have an understanding of the available hardware and its capabilities nor about how the application load translates on load on the physical layers.

Infrastructure models capture the layout of the physical and virtual infrastructure and represent key components such as compute, storage, and network capabilities, as well as their grouping in racks, data centres, and similar. Furthermore, they describe capabilities of the hardware including hardware architecture, virtualisation platform (e.g. type of hypervisor), and virtual machines (containers) running on the host.

Load translation models enhance the infrastructure models and provide a mapping from workload on application components to resource demands on the physical infrastructure. They are crucial for understanding whether enough physical resources are available to handle workload on application level. In addition, they describe the impact of congestion caused by components with similar hardware demands concurrently running on the same hardware.

Finally, **Quality-of-Service (QoS) models** provide a means to express QoS demands towards an application and monitor the fulfilment of these QoS requirements. In addition, they are able to represent the interdependencies between QoS aspects on different levels, e.g. what QoS requirements at the infrastructure level follow from QoS requirements on the application level. QoS models may be taken as constraints for the optimisation problems solved for rearranging application and infrastructures.

1.3.3 Modular Design

A modular architecture is an architecture where at least some components are optional and there exists the ability to add or remove modules or components according to the needs of a given use case (Aissaouii et al. 2013). The benefits of modular design are well known, not least it supports separation of concerns and provides greater implementation flexibility thus reducing costs and risk. A discrete module or component can be implemented without having to implement the entire system. Enterprises, cloud service providers, and cloud carriers (to a lesser extent) come in all sizes and with their own constraints. A modular design provides these firms with greater choice and flexibility.

1.3.4 Machine Learning and AI for IT Operations

As discussed above, the complexity and scale of distributed cloud infrastructure increasingly require an automated approach. As the deluge of data generated by IoT continues to increase, and as demands from new use cases increasingly require edge deployments, e.g. vCDN, the ability of cloud service providers and cloud carriers to respond quickly to demands on infrastructure, service incidents, and improve on key metrics decreases (Masood and Hashmi 2019). Increasingly, enterprises are looking to AI for IT Operations (or AIOps).

AI for IT Operations (AIOps) seeks to use algorithms and machine learning to dramatically improve the monitoring, operation, and maintenance of distributed systems (Cardoso 2019). Although at a nascent stage of development, AIOps has the potential of ensuring QoS and customer satisfaction, boosting engineering productivity, and reducing operational costs (Prasad and Rich 2018; Dang et al. 2019). This is achieved by:

1. automating and enhancing routine IT operations so that expensive and scarce IT staff have more time to focus on high value tasks,
2. predicting and recognising anomalies, serious issues, and outages more quickly and with greater accuracy than humanly possible thereby reducing mean time to detect (MTTD) and increasing mean time to failure (MTTF), and
3. suggesting intelligent remediation that reduces mean time to repair (MTTR) (IBM 2019; Masood and Hashmi 2019).

Predictions suggest that by 2024, 60% of enterprises will have adopted AIOps suggesting that novel solutions to capacity provisioning must accommodate this shift in enterprise IT operations (Gillen et al. 2018).

1.4   Operational Modes

A model-centric approach assumes cloud-edge applications, and the environments that they run in, can be described by a set of models and that, based on these models, it is possible to optimise both cloud-edge infrastructures and their applications at run-time. As such, an optimisation (control) system and mechanism for creating, validating, and extrapolating these models to large-scale environments are required. This requires a variety of interoperating components, which we refer to here as modes.

Data Analytics Mode: The creation of high-quality models requires an in-depth understanding of many aspects ranging from users to application to infrastructure. For deriving this understanding, a sufficient amount of data needs to be available that can either come from a live system or be derived from a simulation environment. The Data Analytics Mode provides the necessary tooling and guidelines to process those data and generate models from it. The analytics itself is a manual or semi-automated process that applies approaches from statistics and machine learning in order to create the models. It consists of pre-processing and data analysis (or model training respectively). When complete, there is a newly generated insight in the form of a mathematical formula, a statistical relationship, some other model, or a trained neural network. These insights form the baseline of the models that are used by other modes and underlying components.

Run-time Operation Mode: The Run-time Operation Mode uses online optimisation to continuously update geo-distributed infrastructure based on the models and the current deployment scenario (deployed applications, available infrastructure, and user behaviour). Data on the actual usage of the hardware and software requirements are collected during run-time. These data are used by optimisers in the system to weight the current placement and usage against other options and come up with new and better configurations. These are output in the form of an optimisation plan that can then be enacted. This changes the configuration of the actual system. The decisions made in order to improve the system are based on mathematical, stochastic, or programmatic models of the system itself, e.g. the capabilities of the hardware, the needs of the application, current and predicted workload in the system, and the movement of users in the real world.

Simulation and Planning Mode: The Simulation and Planning Mode is capable of performing the same steps as the run-time in what-if scenarios and, hence, evaluates the use and acquisition of new, updated, or reallocated hardware. This mode supports scenario (what-if) analyses such as "what if I bought more or different hardware at existing sites", "what if I added a new network site in the topology", and "how much longer can the available hardware handle my workload, if it keeps growing as predicted". Hence, simulation helps operators to take strategic decisions about their infrastructure. What is more, using simulation, different placement scenarios are explored and weighed against each other to serve as calibration and constraints for optimisation algorithms.

1.5   RECAP Conceptual Reference Model

Figure 1.2 presents an overview of the RECAP conceptual reference model which identifies the main components in RECAP and how they interoperate. The diagram depicts a generic high-level architecture and is intended to facilitate the understanding of how RECAP operates.

The diagram below outlines the components in the RECAP architecture and shows the process flow loops in the optimisation framework. The **Landscaper Component (1)** acquires information on the state and configuration of the physical and virtual infrastructure resources from disparate sources and presents same as a graph. The **Monitoring Component (2)** uses probes to collect telemetry metrics needed for the modelling and optimisation tasks, including CPU consumption, disk I/O, memory loads, network loads, and packet statistics—both from virtual and physical resources. These are input to the optimisers and the output is used to orchestrate and enact resource changes in the cloud network.

The **Application Optimiser (3)** is used to optimally autoscale the applications and resources. Application scaling refers to horizontal scaling, namely adding additional application components into the system dynamically, while infrastructure scaling relates to vertical scaling, whereby virtual resources are increased for a component. Applications can be scaled locally or globally and may be in response to run-time traffic limits or

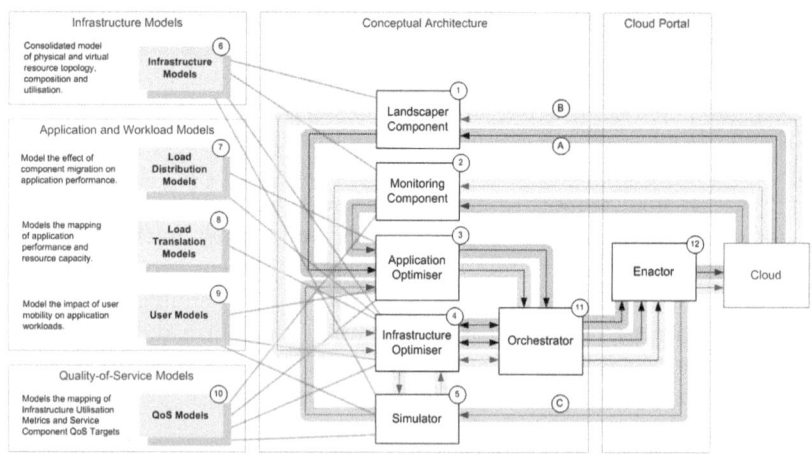

Fig. 1.2   RECAP conceptual reference model

resource levels being reached or may be controlled by data analytic workload predictive systems. The application to be deployed is composed of multiple connected service components in the form of service function chains (SFC), which need to be placed together. In order to achieve better than a very sub-optimal application deployment onto a distributed virtual cloud infrastructure, it is necessary to introduce sufficient functional granularity into the application structure to allow separate components to be provisioned and scaled independently. Application optimisation is essentially a mapping of a graph of application components and dependencies to the network of computing resources that delivers an optimal overall KPI target such as maximum latency or minimum throughput or maximum usage cost. The mapping is done subject to application-specific rules or constraints relating the individual resource requirements for components (Minimum/Maximum instance constraints) and their mutual co-hosting needs (Affinity/Anti-Affinity constraints).

The outputs of the application optimiser are treated as requests or recommendations for application scaling and placement, to be subsequently evaluated by the **Infrastructure Optimiser (4)** which augments the initial placement decision by taking into account the additional knowledge of the available physical infrastructures, the infrastructure policies of the infrastructure provider and specific committed Service Level Agreement (SLA) targets. This allows the infrastructure optimiser to retain full control of the infrastructure resources and to ultimately decide what application requests are enacted and how applications are orchestrated. The **Infrastructure Optimiser (4)** includes (1) Application Placement which optimally maps application components to virtual infrastructure resources to deliver an optimal overall target such as maximum power consumption, maximum operational cost, or specific committed Service Level Agreement (SLA) targets; (2) Infrastructure Optimisation to optimally utilise the physical infrastructure; and (3) Capacity Planning to perform what-if scenarios for additional physical infrastructure.

The Infrastructure Optimiser and Simulator use **Infrastructure Models (landscapes) (6)**. These models/landscapes present the physical and virtual structure, configuration, and topology of the known resources. The telemetry utilisation and performance statistics and the application KPI information are also needed for the Infrastructure Optimiser. Together these inputs form a consolidated infrastructure model that has the appropriate granularity tailored for the given use case thus making optimisation practicably achievable.

Application and Workload Models (7 and 9) describe the application components and their behaviours and dependencies and map the application components with their virtual resource requirements. The **Workload Models** describe the traffic flows through the application components. Both models are used by the workload predictor and application optimiser to forecast workloads and application components and recommend how these components should be placed on the network topology based on optimising the overall application KPIs. The application models describe applications as graphs of components with interdependencies and constraints in the form of graph links. The workload models describe the relationships between control and data plane traffic, between end-to-end latency and traffic, and between traffic and resource usage. They have been built based on the data analysis of historical trace and synthetic workload data using statistical and machine learning techniques.

In the **Application Optimiser (3)**, the traffic workloads are mapped to the application sub-components, and the propagation of workloads is modelled to account for the migratory capability of the components and the mobile nature of users. The Optimisers use **Load Distribution Models (6)** to account for this mobility of application components and the impact of component migration on application performance. They effectively model the traffic flows in the system and can predict the effect on workloads if application components are changed. They are based on the results of load balancing after a component migrates and on user models which drive component migration. These models are used by the optimisers to calculate the cost of component migration when selecting an optimisation option.

Load Translation Models (7) are used by the **Infrastructure Optimiser (4)** to map application configuration to physical infrastructure capacity. The optimiser correlates the virtual resources (VMs/Containers) to physical resources, and the physical resource utilisation with the application component KPIs (throughput, response time, availability, speed of service creation, and speed of service remediation). The translation provides a mapping of actual (specific in time) telemetry metrics of physical resource consumption (utilisation metrics) to application components workloads (i.e. the utilisation of resources by the components that are running on those physical machines). Effectively, this maps the application placement with the performance of components so placed.

The **User Models (9)** are based on an agent-based modelling of users, e.g. citizens navigating through a city and utilising mobile services.

It is possible to create models based on historical trace data and simulated synthetic data. In this case, **Simulators (5)** are a valuable tool for generating the user mobile behaviour and demand for application services as well as the corresponding traffic from the related cloud services.

1.5.1 Optimisation Process Flows

Process A: The **Application Optimiser (3)** is fed with appropriate output from the **Landscaper Component (1)** and **Monitoring Component (2)**, which represents the current resource capacity and utilisation, as well as the **Application Models,** which represent the application workload and performance targets. The **Application Optimiser's (3)** prediction engine produces a recommended deployment of components and outputs this to the **Infrastructure Optimiser (4)** for evaluation, and then to the **Orchestrator (11)** for orchestration. The **Application Optimiser (3)** can be subsequently triggered dynamically to handle variations in application workloads and user behaviours so that placement and autoscaling can take place. In its most proactive mode, the optimiser can create virtual resources, placing and autoscaling based on machine-learning models that are run against workload and user metrics in real-time.

Process B: The **Infrastructure Optimiser (4)** uses the output of the **Landscaper Component (1)** and **Monitoring Component (2),** which represents the current resource capacity and utilisation, as well as the **Workload and Infrastructure Models** to optimise the utilisation of the physical hardware resources based on required Service Level targets and policies. The **Infrastructure Optimiser (4)** optimises the use of the physical resources taking energy, equipment, and operational costs into account as well as the plans and policies around physical resource utilisation. This is based on a logical model of the infrastructure, virtual and physical resources, and their utilisation mappings. The **Infrastructure Optimiser (4)** also needs to represent the mobile nature of workloads and the ability of application component migration to properly optimise the deployment. The Infrastructure Optimiser uses the **Simulator (5)** in a Human-in-the-Loop fashion, using the simulator to formulate deployment mapping selections and calibrating the optimiser's algorithmic process. The **Simulator (5)** validates the results of the optimisation and provides "what-if" scenario planning.

1.6   RECAP BUILDING BLOCKS

While the previous section presents RECAP as a loosely integrated conceptual architecture, this section focuses on four high-level functional building blocks (subsystems) that encapsulate RECAP logic and provide the necessary functionality to realise the three operational modes discussed in Sect. 1.4. The respective building blocks are loosely coupled and are a frame for the RECAP architecture. The building blocks are themselves distributed so that the entire RECAP system represents a distributed architecture. The major functional building blocks (subsystems) are Infrastructure Modelling and Monitoring, Optimisation, Simulation and Planning, and Data Analytics and Machine Learning. Each of the blocks is discussed in-depth in the remaining chapters of the book.

1.6.1   Infrastructure Modelling and Monitoring

The old adage "garbage in, garbage out" particularly applies to making valued optimisation decisions. Thus, within RECAP's Run-time Operation Mode, having an accurate understanding of the current state of applications and the underpinning infrastructure is of paramount importance. Furthermore, the long-term collection of accurate data is a key requirement for being able to apply meaningful data analytics and machine learning strategies (see Data Analytics Mode). Hereby the current state of application and infrastructure is represented by two complementary data sets, the infrastructure landscape and the infrastructure monitoring (telemetry) provided through the Landscaper Component and the Monitoring Component respectively. As discussed earlier, the Landscaper Component is tasked with providing physical and virtual infrastructure data as "a landscape" consisting of nodes and edges. In that landscape, nodes represent for instance physical servers, virtual machines, or application instances. In contrast, edges either represent mappings from applications to virtual resources and further to physical resources, or (network) connections between instances on the same abstraction layer. In short, the Landscaper Component identifies what type of infrastructure is available and where, while the Monitoring Component provides live data from that infrastructure. Both are essential for modelling and optimisation and are encompassed in a requisite distributed design.

As discussed in Sect. 1.5, the RECAP Monitoring Component collects telemetry-like data from physical infrastructure, virtual infrastructure, and applications; stores this data in a unified format; and ultimately provides the data in a consumer-specific format to other components in the wider RECAP system. Both the Landscaper Component and the Monitoring Component have been designed to operate on a per-location (data centre) basis. This helps in respecting administrative domains and, in the case of monitoring, reduces overall network traffic.

1.6.2   Optimisation

Optimisation goals in a multi-tenant distributed cloud-edge environment vary depending on the respective perspective. On the one hand, infrastructure optimisation has the goal to enforce a scheduling strategy that best reflects the intention of the infrastructure provider, e.g. to improve the utilisation of the available hardware or to save energy. On the other hand, application optimisation strategies try to find the best-possible configuration for an application deployment. Hence, the latter will increase the available compute capacity when high workload is expected. This, however, will only lead to satisfaction when the scheduling at the infrastructure level does not apply strategies that counteract these goals. Consequently, RECAP's optimisation subsystem realises a cooperative two-level optimisation framework, in which the optimisers at the two levels (application and infrastructure) interact in order to avoid conflicting scheduling decisions. Besides infrastructure-level and application-level optimisers, the subsystem further contains an optimisation orchestrator that mediates between the two levels. All entities in that subsystem consume monitoring data, application load data, and infrastructure data. The outputs of the optimisation algorithms in turn are optimisation steps that are then processed by the Enactor.

Figure 1.3 illustrates the dependencies between the major components of the optimisation subsystem. While there is just one Infrastructure Optimiser in a given installation, there may be multiple Application Optimisers, one per deployed application. Each of these is equipped with its own application-specific optimisation strategy and optimisation rules. The Infrastructure Optimiser in turn is equipped with provider-specific optimisation policies.

The Application Optimisers constantly receive the current status information from the Infrastructure and Modelling subsystems and, based on this information, estimate the future coming workload. Based on the current and predicted workload, each Application Optimiser suggests

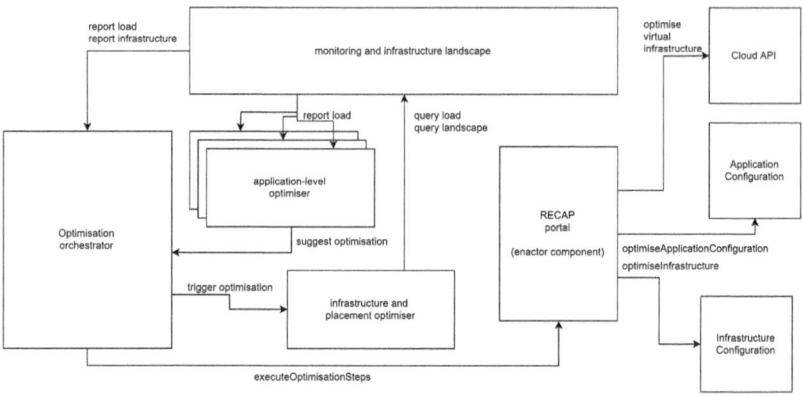

Fig. 1.3 Component-oriented overview of the RECAP optimisation subsystem

optimisation steps for its particular application. These suggestions are fed to the optimisation orchestrator, which, based on the input received, triggers the infrastructure optimiser that then decides on whether these operations are feasible and also the mapping between application components (bundled in virtual machines or containers) and physical resources. Application Optimisation and Infrastructure Optimisation are presented in detail in Chaps. 3 and 4 respectively.

1.6.3   Simulation and Planning

Figure 1.4 illustrates the core architecture of the RECAP Simulation Framework. It consists of an API Component, a Simulation Manager, and Simulation Engines. The API component serves as an entry point for users, be they human or other RECAP components, or external parties. The API Component offers an interface for controlling simulation runs. In particular, it is used for submitting experiments and retrieving simulation results from these runs. From the API Component, the experiment data is forwarded to the Simulation Manager, which, in turn, checks model validity and submits models to an appropriate Simulation Engine. The RECAP Simulation Framework currently supports two simulation engines that address different use case requirements. First, the discrete event simulator (DES), based on CloudSim, is targeted towards the simulation of

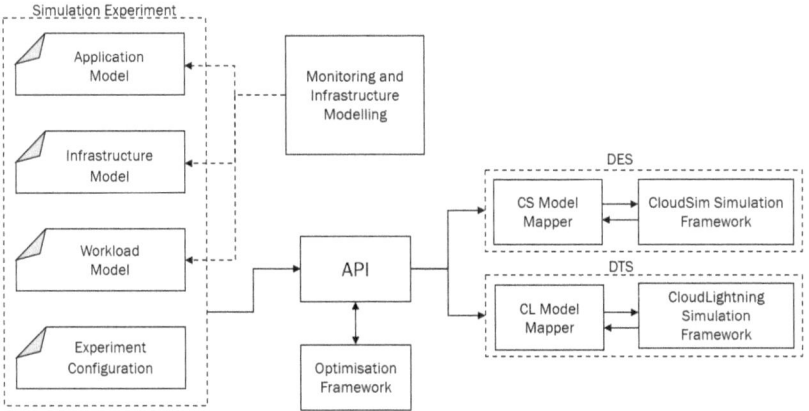

Fig. 1.4 High-level overview on RECAP simulation framework

large-scale cloud-computing infrastructures, data centres, virtual machines, and virtual machine components. It is tailored for fine-grained and detailed simulations. On the other hand, the discrete time simulator (DTS), based on the CloudLightning Simulator, is well suited for large-scale simulations that need to run at speed and whose execution time is bounded.

The primary input to a RECAP simulation is a simulation experiment comprising instances of the application model, the infrastructure model, the workload model, and in addition, an experiment configuration. All of these models are represented in the very same way for both simulation engines. Once the input has been validated by the Simulation Manager, it has to be transformed to the simulation engine-specific format. This is done by the Model Mapper components shown in Fig. 1.4.

1.6.4 Data Analytics and Machine Learning

The Data Analytics and Machine Learning subsystems make use of the data collected by Landscaper Component and the Monitoring Component. The primary goal of this functional block is to distil statistical properties and patterns from load traces. Previously, this activity would be undertaken within an engineering team; however, due to the massive volume of data involved, this can no longer be easily undertaken by humans. As such, the Data Analytics and Machine Learning subsystem operates in a separate processing pipeline that is decoupled from the Optimisation and the

Simulation and Planning subsystems. The steps for analytics cannot be fully automated and require the involvement of a data analyst. Despite this decoupled processing, the results of the analysis do flow back into the RECAP optimisation cycles, either through insights gained by the data analyst performing the analytics (generally in the case of descriptive and/or visual statistical analysis) or through codified models integrated into other RECAP components as libraries or micro-services (more applicable in the machine learning case).

The overall approach of the Data Analytics and Machine Learning subsystem is shown in Fig. 1.5. First, a data scientist retrieves data collected from the Monitoring Component. Then, they perform pre-processing followed by the actual analysis and/or training on the pre-processed data set. Both steps take place in iterations so that the analyst may go back and perform different types of analysis, but they may also go back and perform different types of pre-processing. Finally, as a last step, the results are exported as mathematical models, as codified models, as a library, or as an instantiable service. Due to the decoupled nature of the offline processing, requirements towards the API of the actual data analytics components are less strict than for other RECAP components. The only exception to that rule is the format of the data retrieved from the Monitoring Component. After the data has been fetched, pre-processing and all other steps performed by the data analyst are open and not fixed by APIs. Also, the integration of results into, for example, the optimisation algorithm needs to be defined on a case-to-case basis.

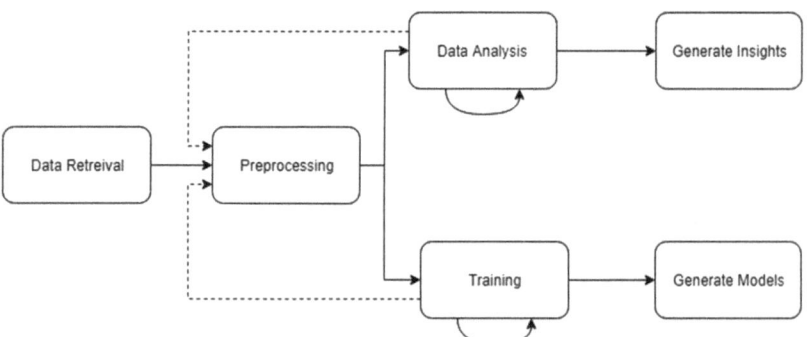

Fig. 1.5 The RECAP approach to retrieve data, analyse it, and export the resulting models to other RECAP components

1.7   Mapping Functional Blocks to Operational Modes

This section describes how the functional building blocks introduced in the previous section interact to deliver the operational modes introduced earlier.

1.7.1   Run-time Operation Mode

The Run-time Operation Mode (see Fig. 1.6) manages a set of applications spread out over a distributed physical and virtual infrastructure such as an IaaS infrastructure with different geo-distributed locations. Based on the user behaviour, and the current and predicted load in the system, the run-time cycle identifies improvements to the current live system on both infrastructure and application level and enacts them by executing optimisation steps. For that purpose, the Run-time Operation Mode makes use of the infrastructure modelling and monitoring subsystem and the optimisation subsystem. Depending on the type of system to optimise, the optimiser may be configured with or without the Infrastructure Optimiser. Not using it yields classical infrastructure unaware application-level optimisation. Internally, the optimisers may make use of additional components generated by the Data Analytics and Machine Learning subsystem. The optimisation plans produced by the optimisers are consumed by the Enactor that interacts with application, physical infrastructure, and virtual infrastructure to enact the optimisations.

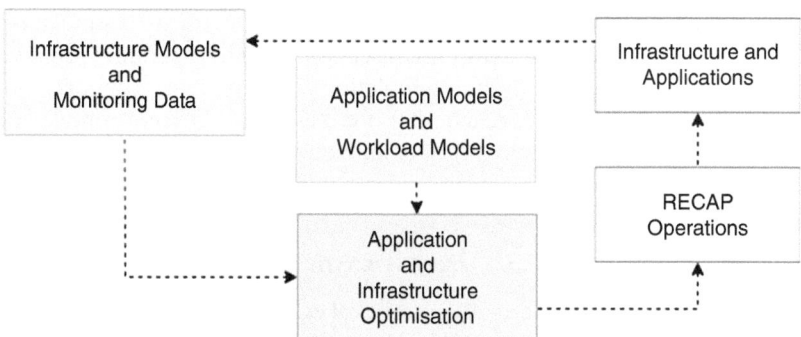

Fig. 1.6   Run-time loop of RECAP

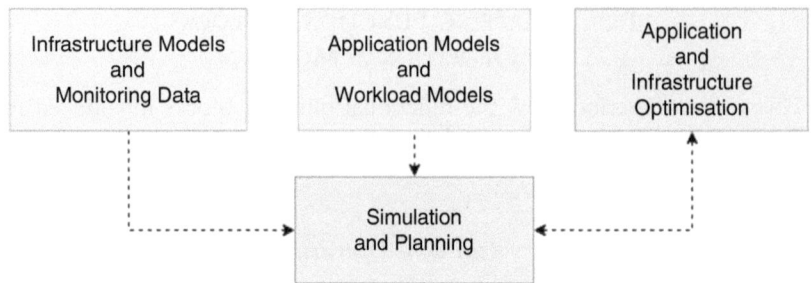

Fig. 1.7   High-level overview on simulation interaction

1.7.2   Simulation and Planning Mode

As discussed in Sect. 1.4, the purpose of the Simulation Mode is to perform two kinds of tasks. Firstly, it helps users and operators conducting experiments about the performance of their infrastructure and applications running therein. This includes the interplay of different types of applications but also the choice of configuration patterns for the Run-time Operation Mode. Secondly, it can be used as a tool for operators to estimate future needs with respect to the amount and type of hardware. Both of these tasks require interaction with the Infrastructure Optimiser.

Figure 1.7 shows how the Simulation Mode is embedded in the wider RECAP architecture. It supports (but does not mandate) importing real-world telemetry and infrastructure landscape data that serve as input to the simulation. These data are combined with the user models, workload models, and load translation models to define a simulation (experiment). Alternatively, parts of the input, or even all of the input, to a simulation can be manually constructed by the user. For helping operators improve their hardware choice, the Simulation Component supports an optimisation-oriented approach that iterates over different simulation configurations and picks the best-possible one for a given application mix and usage scenario.

1.7.3   Data Analytics Mode

The Data Analytics Mode enables statistical evaluation and analysis, as well as applying state-of-the-art machine learning techniques to the data collected by the Monitoring Component. This mode envisions a data

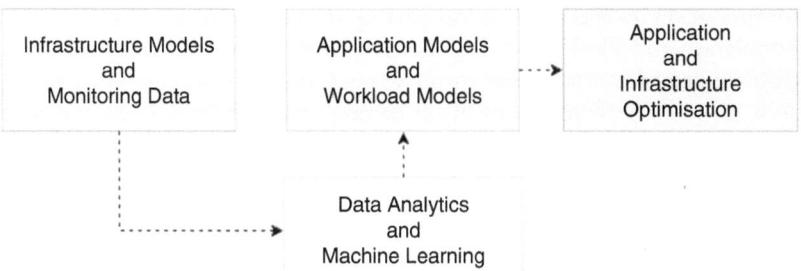

Fig. 1.8 High-level overview on data analytics subsystems

scientist performing many of the steps. Hence, while a certain degree of automation can be achieved in the process, it still requires human interaction, guidance, and input. Figure 1.8 summarises the interaction of the Data Analytics and Machine Learning subsystem with the other RECAP subsystems. It relies on the monitoring subsystems to export metrics as bulk in a normalised manner. This data is then analysed, and the resulting insights and models provided to other RECAP components. In particular, the optimisation components are users of these models, for instance, for the purpose of workload prediction.

1.8   Conclusion

The chapter introduces the challenges of reliable capacity provisioning across the cloud-to-edge continuum. The scale and complexity across this continuum is so complex; it is no longer realistic for IT teams to cost-effectively foresee and manage manually cloud and network operations on a detailed level due to high levels of dynamism and dependencies in the system. This chapter, and the book as a whole, presents a high-level conceptual overview of RECAP—an architectural innovation to support reliable capacity provisioning for distributed clouds— and some of the major design concepts informing its design, namely separation of concerns, model-centricism, modular design, and machine learning and artificial intelligence for IT operations.

The remainder of this book is organised around the four functional building blocks outlined in Sect. 1.6 above. Chapter 2 describes the Data Analytics and Machine Learning subsystem, followed by Application

Optimisation (Chap. 3), Infrastructure Optimisation (Chap. 4), and Simulation and Planning (Chap. 5). The book ends in Chap. 6 with four case studies each illustrating an implementation of one or more RECAP subsystems. The first case study presents a case study on infrastructure optimisation for a 5G network use case. The second case study explores application optimisation for virtual content distribution networks (vCDN) on a large Tier 1 network operator. The third case study presents how data analytics and simulation components, within RECAP, can be used by a small-to-medium-sized enterprise (SME) for cloud capacity planning. The final case study looks at how RECAP components can be embedded in an IoT platform to reduce costs and increase quality of service.

References

Aissaoui, Nabila, Mohammed Aissaoui, and Youssef Jabri. 2013. For a Cloud Computing based Open Source E-Health Solution for Emerging Countries. *International Journal of Computer Applications* 84 (11): 1–6.

Brown, A.W. 2004. Model Driven Architecture: Principles and Practice. *Software and Systems Modeling* 3 (4): 314–327.

Cisco. 2013a. Embracing the Internet of Everything To Capture Your Share of $14.4 Trillion. https://www.cisco.com/c/dam/en_us/about/business-insights/docs/ioe-economy-insights.pdf.

———. 2013b. Internet of Everything: A $4.6 Trillion Public-Sector Opportunity. https://www.cisco.com/c/dam/en_us/about/business-insights/docs/ioe-public-sector-vas-white-paper.pdf.

Cardoso, J. 2019. *The Application of Deep Learning to Intelligent Cloud Operation.* Paper presented at Huawei Planet-scale Intelligent Cloud Operations Summit, Dublin, Ireland.

Dang, Y., Q. Lin, and P. Huang. 2019. *AIOps: Real-World Challenges and Research Innovations.* Proceedings of the 41st International Conference on Software Engineering: Companion Proceedings, 4–5. IEEE Press.

Gillen, A., C. Arend, M. Ballou, L. Carvalho, A. Dayaratna, S. Elliot, M. Fleming, M. Iriya, P. Marston, J. Mercer, G. Mironescu, J. Thomson, and C. Zhang. 2018. *IDC FutureScape: Worldwide Developer and DevOps 2019 Predictions.* IDC.

Haller, S., S. Karnouskos, and C. Schroth. 2009. The Internet of Things in an Enterprise Context. In *Future Internet Symposium*, 14–28. Berlin, Heidelberg: Springer.

IBM. AIOps, IBM Cloud Education. 2019. https://www.ibm.com/cloud/learn/aiops.

Iorga, M., L. Feldman, R. Barton, M.J. Martin, N.S. Goren, and C. Mahmoudi. 2018. Fog Computing Conceptual Model. Special Publication No. 500-325. NIST.

Kleppe, A.G., J. Warmer, J.B. Warmer, and W. Bast. 2003. *MDA Explained: The Model Driven Architecture: Practice and Promise.* Addison-Wesley Professional.

Lynn, T. 2018. Addressing the Complexity of HPC in the Cloud: Emergence, Self-Organisation, Self-Management, and the Separation of Concerns. In *Heterogeneity, High Performance Computing, Self-Organization and the Cloud,* 1–30. Cham: Palgrave Macmillan.

Lynn, T., P. Rosati, and P. Endo. 2018. *Towards the Intelligent Internet of Everything: Observations on Multi-disciplinary Challenges in Intelligent Systems.* Proceedings of the Research Coloquio Doctorados: Tecnología, Ciencia y Cultura: una visión global.

Masood, A., and A. Hashmi. 2019. AIOps: Predictive Analytics & Machine Learning in Operations. In *Cognitive Computing Recipes,* 359–382. Berkeley, CA: Apress.

Papazoglou, M.P. 2012. Cloud Blueprints for Integrating and Managing Cloud Federations. In *Software Service and Application Engineering,* 102–119. Berlin: Springer.

Prasad, P., and C. Rich. 2018. *Market Guide for AIOps Platforms.* Gartner.

CHAPTER 2

RECAP Data Acquisition and Analytics Methodology

*Paolo Casari, Jörg Domaschka, Rafael García Leiva,
Thang Le Duc, Mark Leznik, and Linus Närvä*

Abstract The collection, analysis, and processing of infrastructure information and telemetry data lie at the very heart of RECAP. This chapter describes the infrastructure for the acquisition and processing of data from applications and systems, and explains the methodology used to derive

P. Casari (✉) • R. García Leiva
IMDEA Networks Institute, Madrid, Spain
e-mail: paolo.casari@imdea.org; rafael.garcia@imdea.org

J. Domaschka • M. Leznik
Institute of Information Resource Management, Ulm University, Ulm, Germany
e-mail: joerg.domaschka@uni-ulm.de; mark.leznik@uni-ulm.de

T. Le Duc
Tieto Product Development Services, Umeå, Sweden
e-mail: thang.leduc@tieto.com

L. Närvä
Tieto Sweden Support Services AB, Karlstad, Sweden
e-mail: linus.narva@tieto.com

© The Author(s) 2020   27
T. Lynn et al. (eds.), *Managing Distributed Cloud Applications
and Infrastructure*, Palgrave Studies in Digital Business & Enabling
Technologies, https://doi.org/10.1007/978-3-030-39863-7_2

statistical and machine learning models from this data. These models are then used to identify relevant features and forecast future values, and thus inform run-time planning, decision making, and optimisation support at both the infrastructure and application levels. We conclude the chapter with an overview of RECAP data visualisation approaches.

Keywords Data analytics • Data acquisition • Machine learning • Application modelling • Infrastructure modelling • Distributed cloud computing • Edge computing

2.1 Introduction

The collection, analysis, and processing of data (e.g., infrastructure information and telemetry) lie at the very heart of RECAP and constitute a crucial part of the entire RECAP system. Data make it possible to train machine learning and data analytics algorithms in the analytics mode; moreover, data provide the basis for run-time planning, decision making, and optimisation support at both the infrastructure and the application levels; finally, they can be used as calibration mechanisms for the RECAP simulators. As such, the data acquisition and analytics methodology comprises (1) data acquisition, defining how to collect data from the RECAP infrastructure and the applications running on top of it, how to store that data, and how to provision it to the various parts of the RECAP ecosystem; and (2) data analytics, defining how to access the data and create usable models from it.

Accordingly, this chapter is structured as follows: Section 2.2 describes the infrastructure for the acquisition and processing of data (both from applications and from systems). This is followed by an overview of the data analytics methodology in Sect. 2.3, including the development of mathematical models to identify relevant features and forecast future values. Section 2.4 provides an overview of visualisation in RECAP.

2.2 Data Acquisition and Storage

Data collection in RECAP serves three purposes: (i) to derive information about the flow of messages (hence, the load in the application layer) and use it to create workload and load transition models; (ii) to derive the impact of the application layer behaviour on resource consumption on the

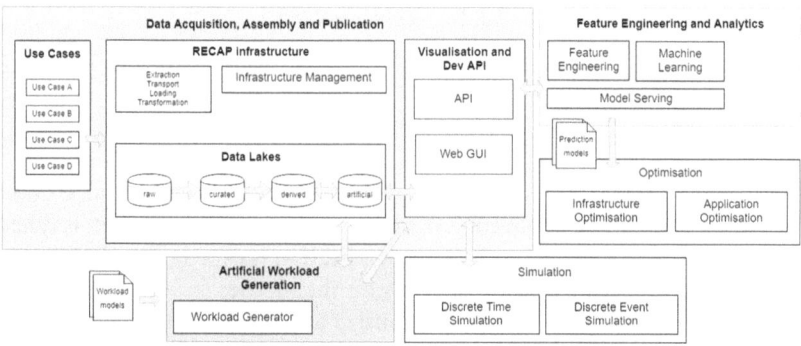

Fig. 2.1 Conceptual overview of data handling in RECAP

physical layer; and (iii) to provide input to simulation and visualisation components.

As shown in Fig. 2.1, RECAP makes use of a central data repository, which serves as the single integration point for all elements of the RECAP ecosystem, and as the primary source of data for other parts of the RECAP platform.

In its databases, the repository stores information about: (i) time series of load metrics, (ii) information about the configuration of the data centre and virtual infrastructure, and (iii) information about the applications running on top of this infrastructure. While (i) is the primary focus of the repository, (ii) and (iii) are additional metadata that enrich the time series data and that help correlate time series of various metrics from different layers of the system. As an example, metadata could help correlate infra-structure metrics, such as CPU usage, with application performance met-rics from the application layer, such as worker queue length.

Technically, the data repository cannot be realised as a single entity, as it has to satisfy different requirements from various components. While the data analytics and machine learning functionality in RECAP require access to large chunks of CSV-formatted data, the visualisation compo-nent requires the capability to flexibly query for data upon a user request. Finally, other RECAP components require access to a live stream of data: for instance, the optimisers constantly need to look up the current state of the system. In consequence, a polyglot approach to persistence is required, as will be presented in later in this chapter.

2.2.1   Terminology

We now briefly cover the terminology that applies to the RECAP Monitoring Architecture.

2.2.1.1 Metrics and Monitoring

Formally, a **metric** is a function that takes a system as input and yields a scalar as a result. The application of a metric on a particular system is called a **measurement** and the result of the application is called the **value** of that metric. The **unit** of the value depends on the metric.

The **monitoring** process continuously (or periodically) applies metrics to systems and generates a series of timestamped values. This is called a **time series** (of a metric).

In order to distinguish values and time series that belong to the same metric, but come from different systems, we allow values to be further enhanced by **metric properties** (or **tags**). This enables values to be grouped, leading to a time series for that tag.

As an example, the cpu_load metric, when applied to a server, yields the current load of the central processing unit on that server. In order to be able to distinguish values measured from server A from those measured from server B, the value may be tagged with the tag origin that in this example can take the values A and B. In total, this creates three time series: one for A, one for B, and one for both servers.

2.2.1.2 Actors

Based on the context of RECAP and the requirements defined by the project's use case providers, the monitoring infrastructure assumes a cloud-like environment where virtual resources (cloud resources) are made available through a Web-based API.

A **(cloud) operator** or **infrastructure provider** provides the physical resources on which virtual resources run. Physical resources may be geographically distributed, leading to a cloud-edge scenario. This actor is responsible for maintaining the physical set-up and for running the software stack that enables access to the virtual resources. The infrastructure provider is also the actor that operates the RECAP infrastructure. Note that communications service providers, such as telecommunications companies, can also be cloud operators and infrastructure providers.

(Cloud) users access the virtual resources offered by the cloud provider. In Infrastructure-as-a-service clouds, they acquire virtual machines

and virtual networks to operate their applications. This makes them
(**application**) **operators** and therefore also users of RECAP.

Finally, **end users** access the applications provided by the application
operator. Usually, they do not care where the application runs, as long as
it provides an acceptable quality of service and experience.

2.2.2 Monitoring Layers

Figure 2.2 illustrates the four layers that can be monitored in order to
derive insights on application behaviour and load propagation. Not all layers are required for all installations, so the set-up presented here is a superset of the possible set-ups.

The **physical layer** is provided by the infrastructure provider, and contains the hardware used to run all higher layers. Here, monitoring metrics
mainly include CPU, RAM, disk, and network consumption at specific
points in time. The layout of the physical infrastructure is also important,
e.g. which servers share the same network storage or uplink to the Internet.
Figure 2.3 shows two data centre locations on the left and right hand
sides, each with a router. Both are connected through the Internet. The
infrastructure provides RECAP-aware monitoring support for the physical
layer and report measurements for the metrics.

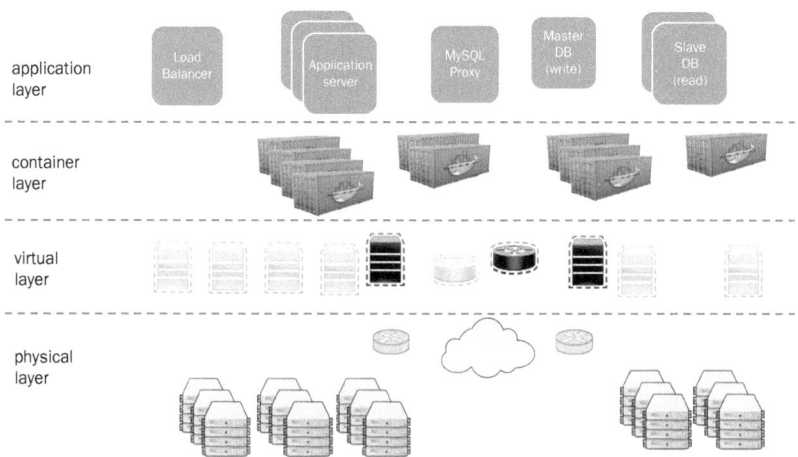

Fig. 2.2 RECAP monitoring layers

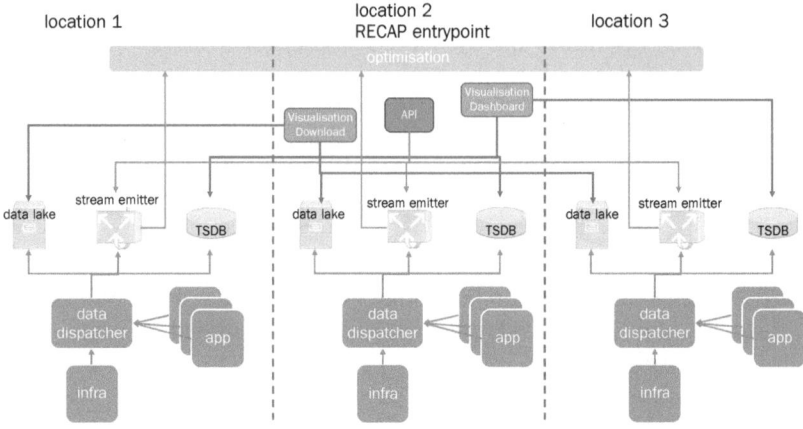

Fig. 2.3 RECAP's distributed monitoring architecture

The **virtual layer** constitutes virtual infrastructure as realised for instance by Infrastructure-as-a-Service (IaaS) clouds. This infrastructure is composed of virtual machines, virtual storage, and virtual networks. It is run by the application operator, which is responsible for the measurements on that layer as well. Similar to the physical layer, metrics mainly include CPU, RAM, disk, and network consumption, but several instances of virtual infrastructure (to be monitored separately) can exist in the virtual layer (cf. different colours in Fig. 2.2). Also, the number of virtual components per virtual infrastructure as well as the number of virtual infrastructures is not fixed, but can dynamically grow and shrink.

On top of the virtual (or physical) layer resides an optional **container layer** such as a Kubernetes[1] cluster or a Rancher Cattle[2] cluster. Basically, the same restrictions and considerations hold for this layer as for the virtual layer. Yet, in contrast to both, the container layer can provide a seamless abstraction and hide the location of different data centres. Whether containers are used is a design choice by application owners or cloud providers: containers can be offered by a cloud provider or be deployed by a user on top of virtual machines.

---

[1] https://kubernetes.io/
[2] https://github.com/rancher/cattle

At the top resides the **application layer** where application- and component-specific metrics can be applied. These include, for instance, the queue length of load balancers, detailed statistics on the use of databases, and the message throughput of a publish-subscribe system. Application-specific metrics are important conveyors of KPIs or QoS. Although the RECAP monitoring platform cannot define all possible application-level metrics to be captured, it provides a structure to measure and store application-level metrics.

2.2.3   *Monitoring Architecture*

The RECAP Monitoring Architecture collects and provides the monitoring data from the four layers described earlier to the RECAP simulator, run-time system, and users.

RECAP operators may manage infrastructure spread over several, geographically distributed locations. In each of these sites an edge or core/cloud data centre resides. In order to limit data hauling across data centres, collected data are stored as close to their origin as possible. RECAP's acquisition and retrieval strategy takes these circumstances into account. In the following, we first describe the acquisition and storage architecture per site and then the overall architecture spanning different sites.

2.2.3.1 *Single Site Monitoring Set-up*

This section describes the monitoring set-up for each site in a RECAP managed infrastructure. Each of the sites can run in isolation and is not affected by traffic and load on other sites. The per-site architecture consists of monitoring probes on the physical layer and on the higher layers of the software stack. It also involves a data dispatcher that filters the incoming data and relays it to three different data sinks: a data lake, a time series database, and a data stream emitter.

Components

Probe: Probes convey monitoring data out of monitored systems. Different probes may be necessary for different metrics, even though most probes will perform measurements for multiple metrics. Each probe may emit data in a different format and at different time intervals. In addition to a timestamp and value, an emitted data item contains metric properties to identify the source and scope of the data point. While it is the responsibility of the infrastructure operator to provide probes for the physical layer, the cloud user shall provide the necessary metrics for higher layers of the stack. All probes directly or indirectly send their data to the data dispatcher.

Data dispatcher: The data dispatcher is a site's central data monitoring integration point. It receives and normalises all probe data and sends them to the data sinks. Normalisation depends on the probes: an individual transformation is needed per metric and per probe type. The dispatcher adds site information and similar attributes to collected data.

The data normalised by the dispatcher are then put in the sinks. In RECAP, different data post-processing demands exist with regard to the monitoring subsystem. Feature engineering and data analytics in RECAP operate on large data sets which need to be processed offline in dedicated servers. Visualisation works on smaller datasets, but requires high flexibility in data provisioning. Finally, optimisers require a snapshot that represents the most recent state of the managed infrastructure. Therefore, RECAP applies three different types of sinks:

- **Data lake sink**: accumulates large amounts of monitoring data in a durable storage for a long time using a compact representation. This data is the basis for data analytics and machine learning.
- **Time series database sink (TSDB)**: stores monitoring data in time series. Through an underlying indexed search engine, it supports live queries of current and past data. It is the primary data source for the visualisation components.
- **Stream emitter sink**: relays a configurable subset of live monitoring data to other parts of the RECAP infrastructure. It is the primary data source for the application optimisation and infrastructure optimisation engines, which decide on the metrics of interest and any pre-processing (e.g. smoothing) to be applied.

A distinctive trait of RECAP is the "separation of concerns" between application and infrastructure optimisation procedures. This makes it possible to accommodate the (often contrasting) objectives, costs and constraints of both application and infrastructure providers, and to harmonise them as far as possible within the RECAP framework.

Practical Considerations

The dispatcher and all sinks are stateless and can be scaled to serve large hardware installations, large amounts of users, and high volumes of data.

Neither the architecture nor its implementation puts any restrictions on where dispatcher and sinks can be run. Yet, in order to ensure a correct interplay of the acquisition and storage components with other parts of the RECAP infrastructure, the following constraints have to be considered:

1. All kinds of probes at a site need to be able to connect to the dispatcher of that site either directly or indirectly.
2. Both the TSDB and data lake sink need to be accessible from the API component described later in Sect. 2.3.2.
3. The stream emitter sink needs to communicate to other data centres, and particularly to the optimisation subsystem.

2.2.3.2 Cross-site Monitoring Set-up

As RECAP provides cross data centre resource and application management, individual sites must be spanned to achieve a holistic view of the system. The Monitoring Architecture achieves this by introducing a RECAP entrypoint that may also be bound to a DNS name in order to ease access, and includes a load balancer to point to the various sites managed by this instance of RECAP.

Figure 2.3 provides an overview of the overall architecture of the monitoring infrastructure spanning sites. It shows three locations, one of which functions as the RECAP entrypoint. Besides the local entities from Sect. 2.3.1, it shows the visualisation endpoints that offer a dashboard with usage graphs as well as a GUI for bulk download of data from the data lakes. The more generic API entity component serves as an integration point for other RECAP components. In particular, the optimiser can use it to configure the stream emitter sink which provides input to the optimisation cycle or in order to access time series data from the TSDB.

As detailed earlier, the data lake sink is instantiated per site and can be a distributed component that compresses and stores raw monitoring data. Its primary purpose is to serve files for bulk download. As this storage form is resource hungry, the monitoring infrastructure (1) switches off persisting raw metrics on a per-site basis (this is beneficial if the site cannot store larger amounts of data or no later data analysis shall be performed), and (2) deletes or moves away data older than a certain age. While this creates cross-site load, the fact that data is sent filtered and compressed requires much less bandwidth than uncompressed probe data.

2.2.4   Data Structure for Storage

This section introduces the actual data that is collected on the four layers. We do not discuss the data sent by the various probes as RECAP does not enforce the use of specific probes. Instead, it assumes that the dispatcher performs probe-specific normalisation.

2.2.4.1   Metrics on the Physical Layer

The metrics gathered for the physical layer are split into seven metric sets (cpu, diskio, filesystem, memory, vms, and vm). Each of the metric sets contain several detailed metrics. The four metric sets host.cpu, host.diskio, host.memory, and host.filesystem capture the detailed usage and utilisation of basic system resources (cpu, block devices, memory, file systems). Instead, host.vms gives information about the virtual machines running on a host, and the metrics from the host.vm metric set detail the resource consumption per virtual machine.

We measure the resource consumption of a virtual machine from the host to avoid the misinterpretation of numbers seen from inside the virtual infrastructure. For example, a 100% CPU load seen inside the virtual machine may not mean that the machine uses a full physical core. The mapping of how many physical cores are represented by one virtual core for this particular virtual machine is subject to the CPU scheduler on the host/hypervisor and is heavily influenced by the overbooking factor of the physical server. Hence, the physical layer needs to report on the physical resource usage per virtual machine.

2.2.4.2   Metrics on the Virtual Layer

The metrics gathered at the virtual machine level (i.e. captured from within a virtual machine) start with vm. and are basically the same as the physical host except, for example, cpu.steal. In addition to resource consumption, information about available containers is collected in the same way as resource utilisation per container.

2.2.4.3   Metrics on the Container Layer

On the container level, we collect the very same metric sets and metrics as for the virtual layer (cpu, diskio, memory, filesystem, and network). The names of the metric sets start with container. instead of vm.

2.2.4.4 Metrics on the Application Layer

Applications differ and so do the metrics that can and need to be collected from them. In particular, the measurement gathering methods depend on the application and its software components. Hence, the data format and content for application metrics cannot be fixed in advance, and metric collection must be part of the application lifecycle management.

A generic naming convention for application-level metrics is adopted in RECAP with the format app.<app name>.<comp name>.<metric name>, which includes the (system-wide unique) application name, the component name (unique per application), and the metric name.

2.2.4.5 Metric Attributes: Tagging

So far, we have presented metrics per layer. Yet, with the information provided so far, it is not possible to distinguish data from different sources. This is achieved via metric attributes that also enable data grouping and correlation. For example, all metrics are tagged with the timestamp and the layer (physical, virtual, container, application). All physical layer metrics are further tagged with the data centre location, the name of the physical host, and the name of the infrastructure provider. Metrics on the virtual layer are enriched with information about the cloud they are running in, the current region they reside in, and their respective identifier. Similarly, container metrics contain information about the container identifier. On all levels, specific attributes are added if required by the metric. For instance, devicename helps distinguish network interfaces on physical hosts.

For application metrics, tagging needs to fulfil two orthogonal tasks: to distinguish different instances of the same application (e.g. WordPress installation for customer A and customer B), and to distinguish different instances of an application component, e.g. a scaled out application server. Hence, all application metrics are tagged with an application instance identifier and a component instance identifier, both automatically assigned by the platform and added by the RECAP data dispatcher system. If needed, application owners can provide further tags.

Second, tagging needs to convey on what physical resource an application or component was running. Therefore, all application metrics are tagged with the type and identifier of the containing entity (e.g. virtual machine or container).

2.2.5   *Implementation Technology*

The implementation technology for the monitoring system chosen for RECAP is largely based on experience gained from the FP7 CACTOS project (Groenda et al. 2016) using an OpenStack testbed and production system (bwCloud[3]) and from the Horizon 2020 Melodic project (Melodic 2019). Where possible, all technical building blocks were components where technology was available under an open source and/or a commercial licence. Finally, no chosen component makes any assumptions on the technology of the other components, facilitating replacements and upgrades.

The **data dispatcher** is realised through Elastic Logstash[4] which offers pipelines for receiving, processing, and dispatching a wide range of monitoring data. It comes with an extensive list of input plugins, including software to accept TCP/UDP network traffic with JSON payload. Output plugins range from time series databases and overwrites to the file system to sending message streams through publish-subscribe platforms such as Apache Kafka. Filters are provided for data curation and transformation.

The **time series database sink** is realised via an InfluxDB instance, which supports both groups of metrics and metric attributes/tags. It also supports continuous queries and data aggregation, and integrates well with Grafana, an open source metric analytics and visualisation suite commonly used for visualising time series data for infrastructure and application analytics.[5]

The **stream emitter sink** is realised by the Apache Kafka[6] publish-subscribe system, due to its wide adoption and well-known scalability.

The **data lake sink** is based on CSV files stored in a compressed format.

Probes: The entire monitoring subsystem is independent from the specific monitoring technology. This allows RECAP to integrate into existing installations. Consequently, running RECAP does not require operators to perform major updates on their infrastructure. Therefore, the mapping from the data collected by the probes to the metrics schema must be implemented for the dispatcher per probe type. Based on the RECAP testbeds, a set of mapping rules have been implemented for specific probes. In particular, this is the case for the Elastic Metric Beat metric collector to

---

[3] https://www.bw-cloud.org/
[4] https://www.elastic.co/products/logstash
[5] https://grafana.com
[6] https://kafka.apache.org/

collect metrics on the physical and virtual layer, for Intel's SNAP collector to collect metrics on the virtual container and application layer, and for a VMware vSphere[7] collector.[8]

2.3   Data Analytics and Modelling

2.3.1   Data Analytics Methodology

In this section, we describe the RECAP methodology for the analysis of datasets and the development of machine learning algorithms to support the application of RECAP's results to new problems related to optimal resource allocation and capacity planning. The methodology is composed of five main steps as outlined in Fig. 2.4.

2.3.1.1   Step 1: Problem Definition and Data Assembling
The initial steps are to *identify the problem* to be solved and *the available data* that can help solve the problem through machine learning. In RECAP we merged these steps into a single task due to their high interdependence. If the available datasets are insufficient, we have to change our expectations about the problem or find additional data. As an alternative, we later explain how to enrich existing datasets with synthetic datasets mimicking the same workload data collected from RECAP Use Cases.

2.3.1.2   Step 2: Metric for the Evaluation of the Results
Selecting the metric to evaluate the results of our model is critical, since that metric is exactly what the training algorithm will optimise. If the output of the model is a continuous variable, the Root Mean-Square Error (RMSE) is a typical choice. In the case of a categorical response, typical metrics are accuracy, or the area under the receiver operating characteristic (ROC) curve (AUC).

There are multiple standard techniques to evaluate the performance of a machine learning model and detect issues, such as overfitting, early. These include train/test splits of the dataset, N-fold cross-validation, and bootstrapping. In RECAP, we use train/test splits for the early model prototyping, and apply a cross-validation to the final models before

---

[7] http://www.virten.net/2015/05/vsphere-6-0-performance-counter-description/
[8] https://github.com/Oxalide/vsphere-influxdb-go

Fig. 2.4 A summary of the main steps of the methodology for exploratory data analysis of new datasets

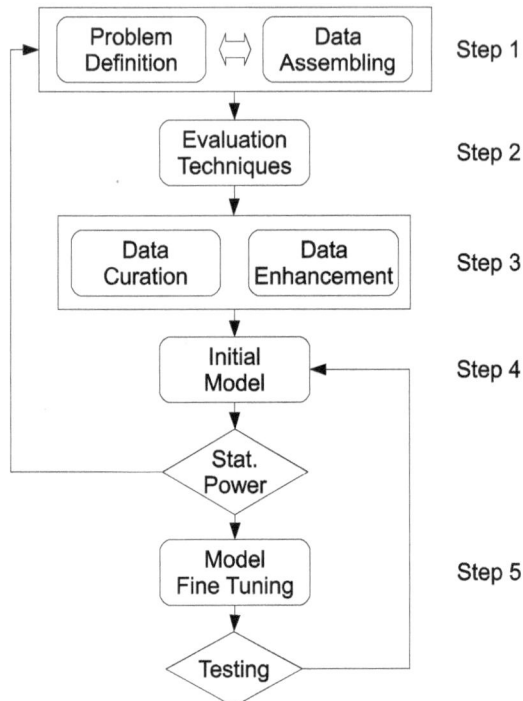

production. Techniques to avoid cross-validation altogether have also been investigated as a promising research direction (García Leiva et al. 2019).

2.3.1.3 Step 3: Data Curation and Enhancement

A data curation process to remove errors and anomalies and fix missing data is an important preparatory step before training a model. A visualisation of the dataset and a descriptive analysis provides valuable information about the quality of the data being used in the project. Outlier detection or the identification of 'Not Available' values could be applied as well. It might also be necessary to enhance the data by deriving new features based on those that already exist. This data enrichment could significantly improve the predictive capabilities of models.

2.3.1.4 Step 4: Model Development

Identifying the best model is often a daunting task in the presence of all possible alternatives. For example, in the case of a classification problem, we could apply techniques like K-means, decision trees, support vector machines, or neural networks. Moreover, each technique could have different alternative configurations. An approach to speed up the selection of the right family of models is to test the statistical power of the machine learning techniques. This test consists of performing fast training of the model, perhaps with a data subset, and in checking if the model has better predictive capabilities than random guessing. Any family of models with no predictive power should be discarded.

2.3.1.5 Step 5: Regularisation and Hyperparameter Selection

The final step of the methodology is to tune the model's hyperparameters, whose values must be set before the learning process begins. Hyperparameter optimisation makes it possible to obtain the best predictive capabilities from a machine learning model, at the price of a higher risk of overfitting. Once hyperparameters have been optimised, the model can be applied to test data never used during training and validation. A clear sign of overfitting is then a divergence between test performance and validation performance.

2.3.2 Exploratory Data Analysis

Descriptive statistics are metrics that quantitatively describe, characterise, and summarise the features of a data set. Even when data analysis draws its main conclusions using inferential statistics and predictive analytics, descriptive statistics can be used to provide a summary of the types of data involved in the use cases, and inform future inference and prediction steps.

Exploratory data analysis (EDA) is used to understand data beyond formal modelling or hypothesis testing. EDA is useful to check assumptions required for model fitting, to handle missing values, and understand the required variable transformations. Figure 2.5 shows an example of a decomposition of a time series in order to visually identify trends and possible cycles. The top panel visualises the original time series. From this data, we extract a trend (second panel), a seasonal component showing clear cyclic behaviour (third panel), and a residual behaviour not explained by trend and seasonal components (bottom panel). These exploratory steps are helpful to inform the choice of time series prediction techniques.

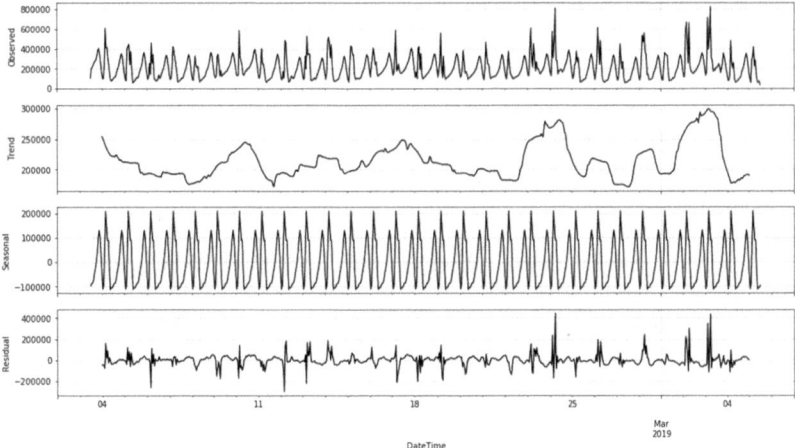

Fig. 2.5 Decomposition of received traffic at a cache

2.3.3   *Workload Prediction*

After a careful survey of the available literature in the field (Le Duc et al. 2019), three techniques were considered for the specific task of workload prediction—probabilistic models, regression-based models, and machine learning models.

2.3.3.1 Probabilistic Models

Probabilistic models are powerful tools to explain datasets, and are widely used in statistics, traffic engineering, simulations, etc. To facilitate workload prediction in RECAP, we attempt to fit several probability density functions to our datasets on a per-use-case basis. Parameter fitting is obtained through Maximum-Likelihood Estimation, and the resulting models are compared through the Kolmogorov-Smirnov test. The best fitting model is finally chosen (an example for cache content pulling is provided in Fig. 2.6).

2.3.3.2 Regression-based Models

Regression-based models are often simple and robust in generating predictions, and thus particularly suitable for offline modelling and prediction tasks. In RECAP, we consider autoregressive integrated moving average (ARIMA) models, which are composed of three parts. The AR part relies

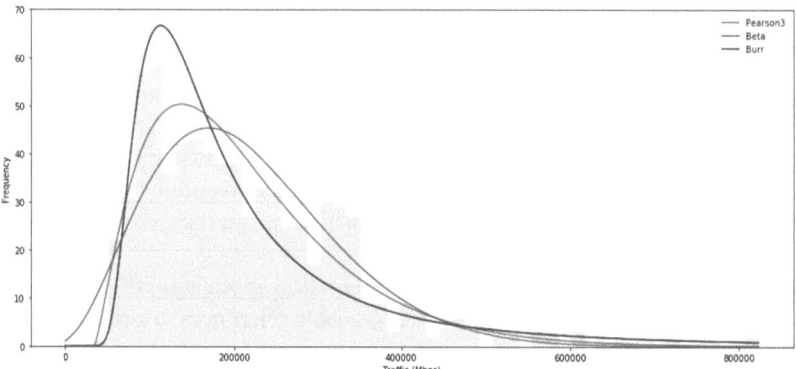

Fig. 2.6 Statistical distributions fitted to records of data sizes of pulled cache content

on the lagged values of the variable of interest; the MA part is actually a linear combination of error terms whose values occurred in the past; and the I part (for "integrated") indicates that the data values have been replaced with the difference between their values and previous values. We also extend ARIMA models with seasonal components (SARIMA).

2.3.3.3 Machine-Learning-based Models

In order to facilitate fast online workload predictions in RECAP, we consider the Online Sequential Extreme Learning Machine (OS-ELM), which enables the generation of workload models and predictions online, and can flexibly handle workload changes. OS-ELM is an efficient technique for online time series modelling and prediction due to its accuracy comparable to batch training methods and to its extremely fast generation of predictions (Huang et al. 2005; Liang and Huang 2006). It accepts input data either sample-by-sample or through varying- or fixed-size data chunks.

Different from other learning methods (e.g. single hidden layer feed-forward neural networks), OS-ELM randomly initialises input weights and updates output weights using the recursive least squares method. This makes OS-ELM adapt quickly to new input patterns, and results into a better prediction performance than other online learning algorithms (Park and Kim 2017).

2.3.4   *Artificial Workload Generation*

RECAP is interested in the availability of datasets describing the evolution of workloads of servers and services (applications), so that stochastic models can be trained to forecast future workloads. However, for reasons including commercial sensitivity and privacy, such datasets may be insufficient for research tasks. This issue can be circumvented by generating synthetic datasets that preserve the statistical properties of the datasets collected from real infrastructures.

Here, we briefly introduce the mathematical models used to generate artificial workload traces in RECAP. Relevant references are provided for the interested reader.

2.3.4.1 *Structural Models-based Workload Generation*

Structural time series models are a family of stochastic models for time series that includes and generalises modelling techniques, including ARIMA or SARIMA models (Harvey 1989). A structural time series model expresses an observed time series as the sum of simpler components:

$$f(t) = f_1(t) + f_2(t) + \ldots + f_n(t) + \epsilon$$

where ϵ is a white error term following a normal distribution of mean 0 and variance σ^2.

For example, one component might encode a linear trend, a cycle, or a dependence of previous values. Structural time series models identify and encode assumptions about the processes that have generated the original data. In this way, they make it possible to generate artificial data traces that have the same statistical properties as the original datasets. The application of a structural time series model to requests coming to a search engine web server is shown in Fig. 2.7. We observe that predicted data (light grey) mimic well the general characteristics of ground truth (dark grey).

2.3.4.2 *GAN-based Workload Generation*

Synthetic data generation using Generative Adversarial Networks (GAN) has recently gained popularity. A GAN is based on a combination of two neural networks, a discriminator (D) and a competing generator network (G). In the training phase, D is trained to distinguish real data from generated data. In parallel, G is trained to fool D by producing better and better fake data that D will eventually accept.

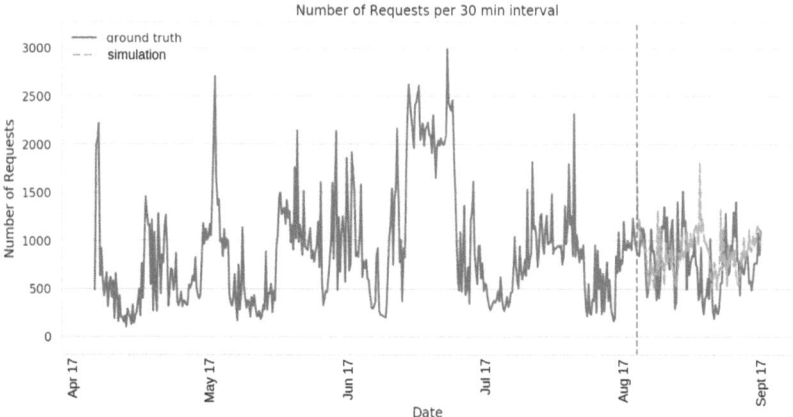

Fig. 2.7 Simulated workload for a search engine

In RECAP, the overall idea behind the use of GANs is twofold. Firstly, using this approach provides a "what-if" analysis on a dataset, answering such questions as "how would this workload look for a larger number of nodes?" Secondly, the inherent training goal of a GAN is to estimate the probability distribution of the training data and to generate synthetic samples drawn from that distribution. Hence, when applied to a real dataset, the GAN learns to mimic its statistical properties.

2.3.4.3 Traffic-Propagation-based Workload Generation

RECAP implements five diffusion algorithms for workload generation. These algorithms can be divided into two groups: non-hierarchical and hierarchical workload diffusion. The former includes population-based, location-based, and bandwidth-based algorithm; the latter includes hierarchy-based and network-routing-based algorithm.

Diffusion algorithms can be applied in different use cases and under different assumptions related to the network topology, network links' capacity, and the distribution of users throughout the network. Given these models, and real workload data traces collected as time series at a limited number of locations, it is possible to produce workload traces for any or all network locations.

2.3.4.4 Simulation System Model Data Sets

The role of simulation in the RECAP project is to go beyond the limitations of an available testbed in terms of scale and complexity of experimentations. Based on simulation, it is possible to generate synthetic datasets consisting of two parts: large-scale models of a system that is being simulated, and simulated behaviour measurements of the modelled system. This is discussed further in Chap. 5.

2.4   DATA VISUALISATION

Data visualisation empowers end users and data scientists to analyse and reason about data and its features. With data visualisation, data sets produced by RECAP or collected from production systems of use cases are transformed to be more accessible, understandable and consumable. RECAP uses a range of visualisation tools which we will now discuss.

2.4.1   Visualisation for Data Analysis

To facilitate data analysis and reasoning, RECAP has adopted various visualisation tools for data presentation, for instance the histogram, box plot, and scatter plot. Upon dealing with heterogeneous data sets, the selected tools enable both univariate and multivariate data visualisation, facilitating corresponding data analysis methods applied to different data sets. To illustrate the use of the visualisation tools as well as their facilitation of data analysis, different visualisations of features extracted from a real data set from a search engine are provided along with explanations of how each visualisation helps retrieving insights into the data.

Figure 2.8 visualises univariate data (specifically, the serving time of user requests in the given workload data set) in different forms. The histogram provides the insight that the majority of user requests are served within very short time periods. The observable data distribution suggests a potential application of probabilistic modelling techniques is needed to construct models of the feature for further analysis or workload generation. The box plot of this serving time feature shows a large number of outliers exist in the data set. Further investigation is thus required for hints on the construction of predictive models. Figure 2.9 visualises multivariate data and shows a relationship between the response size and response time of the user requests. This visualisation suggests a correlation analysis on the data set is needed when addressing workload analysis and modelling.

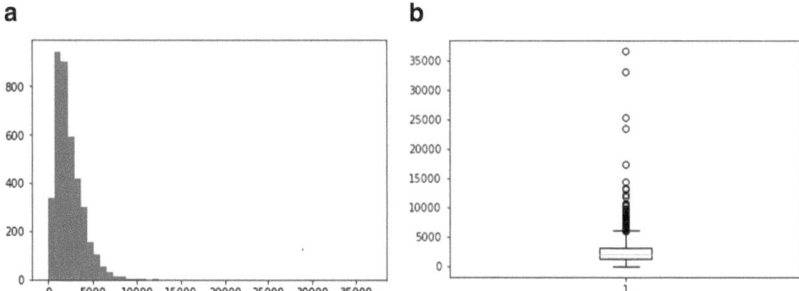

Fig. 2.8 An exemplary presentation of serving time of requests in a workload data set. (**a**) Histogram of serving time of user requests. (**b**) Box plot of serving time of user requests

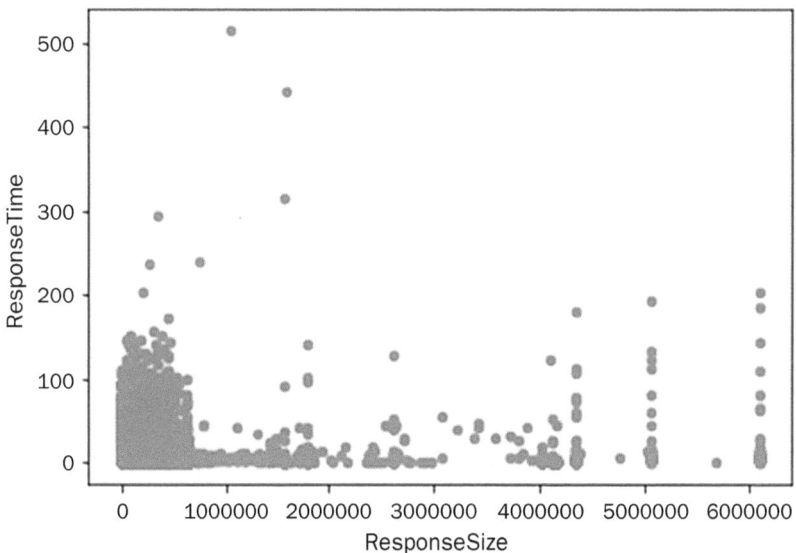

Fig. 2.9 An exemplary presentation of a correlation of features in a workload data set

2.4.2 Visualisation of RECAP Telemetry Data

The visualisation of telemetry data makes the status of the infrastructure and applications operating on the infrastructure more comprehensible for the operators at both application and infrastructure level. This becomes crucial for the automation of system (application and infrastructure) management, in which trust is required and can be established based on visualisations illustrating the response of the system to the triggered and ongoing management actions. In RECAP, telemetry data acquired from use case testbeds and production systems need to be visualised in order to aid the analysis of the workload and application behaviours as well as the mutual dependencies between metrics or features of both the infrastructure and applications.

As discussed, to facilitate the visualisation, Grafana was used as a visualisation tool. This is an open source tool with a large community and a wide selection of plugins and pre-configured dashboards which accelerates visualisation. Grafana has an easy-to-use interface with various graph visualisation techniques including line graphs, bars, heat maps, maps, and architecture. It enables grouping various graphs into a single-view dashboard and supports multiple dashboards to provide different perspectives of a given data set. Figures 2.10 and 2.11 illustrate the snapshots of two dashboards. The first includes multiple graphs showing resource utilisation of the core of a testbed deployed at Ulm University (UULM), Germany, and the second illustrates the mobility and behaviour of users emulated in a testbed deployed at Tieto, Sweden, in a study of Infrastructure and Network Management.

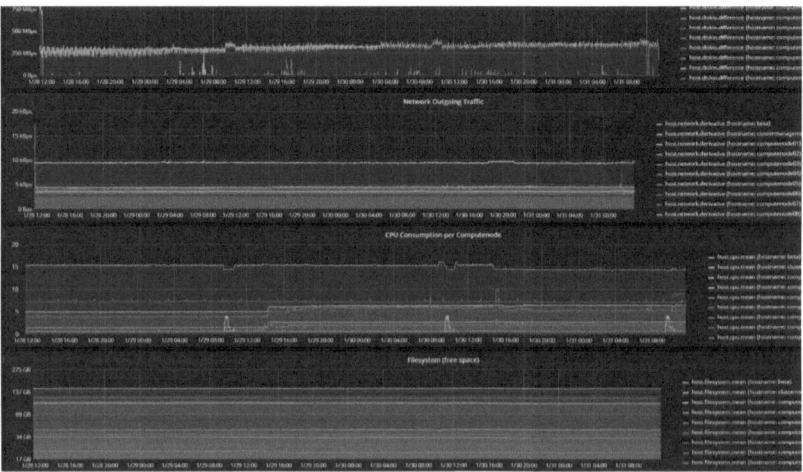

Fig. 2.10   Snapshot of the dashboard for the testbed at UULM

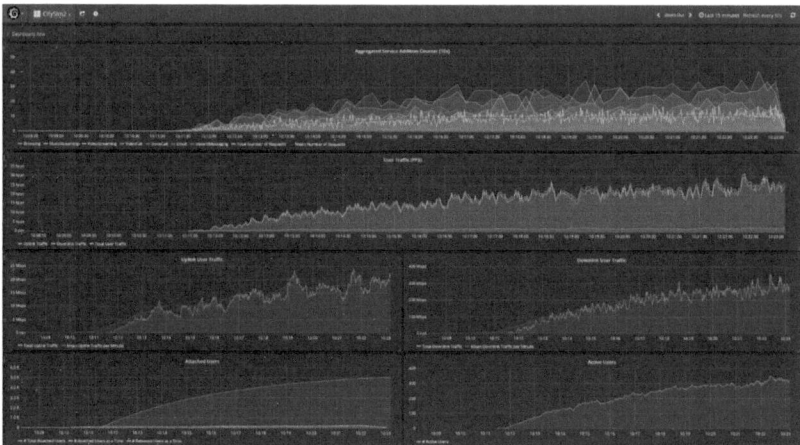

Fig. 2.11 Snapshot of the dashboard for the testbed at Tieto

2.5   OPEN DATA

The RECAP project adheres to the Open Data Pilot of the European Commission. This means that the project committed to providing the datasets required to reproduce the results in the project, unless this would result in, for example, a breach of confidentiality for the dataset provider or in the loss of intellectual property. Several datasets have been derived and provided in the context of RECAP. These datasets are described in RECAP's Deliverable D5.3 and are available, where appropriate, at RECAP's website—https://recap-project.eu.

REFERENCES

Le Duc, Thang, Rafael García Leiva, Paolo Casari, and Per-Olov Östberg. 2019. Machine Learning Methods for Reliable Resource Provisioning in Edge-Cloud Computing: A Survey. *ACM Computing Surveys* 52 (5): 94:1–94:39.

García Leiva, Rafael, Antonio Fernández Anta, Vincenzo Mancuso, and Paolo Casari. 2019. A Novel Hyperparameter-Free Approach to Decision Tree Construction That Avoids Overfitting by Design. *IEEE Access* 7: 99978–99987.

Groenda, Henning, et al. 2016. CACTOS Toolkit Version 2. CACTOS Project Deliverable.

Harvey, A.C. 1989. *Forecasting, Structural Time Series Models and the Kalman Filter*. Cambridge University Press.

Huang, Liang, et al. 2005. *On-Line Sequential Extreme Learning Machine.* The International Conference on Computational Intelligence.

Liang, N.-Y., and G.-B. Huang. 2006. A Fast and Accurate Online Sequential Learning Algorithm for Feedforward Networks. *IEEE Transactions on Neural Networks* 17 (6): 1411–1423.

MELODIC. 2019. *Multi-cloud Management Platform.* http://www.melodic.cloud/.

Park, Jin-Man, and Jong-Hwan Kim. 2017. *Online Recurrent Extreme Learning Machine and Its Application to Time-Series Prediction.* Proceedings of the International Joint Conference on Neural Networks (IJCNN), Anchorage, AK.

Application Optimisation: Workload Prediction and Autonomous Autoscaling of Distributed Cloud Applications

Per-Olov Östberg, Thang Le Duc, Paolo Casari, Rafael García Leiva, Antonio Fernández Anta, and Jörg Domaschka

Abstract Optimisation of (the configuration and deployment of) distributed cloud applications is a complex problem that requires understanding factors such as infrastructure and application topologies, workload arrival and propagation patterns, and the predictability and variations of user

P.-O. Östberg (✉)
Umeå University, Umeå, Sweden
e-mail: p-o@cs.umu.se

T. Le Duc
Tieto Product Development Services, Umeå, Sweden
e-mail: thang.leduc@tieto.com

P. Casari • R. García Leiva • A. Fernández Anta
IMDEA Networks Institute, Madrid, Spain
e-mail: paolo.casari@imdea.org; rafael.garcia@imdea.org;
antonio.fernandez@imdea.org

T. Lynn et al. (eds.), *Managing Distributed Cloud Applications and Infrastructure*, Palgrave Studies in Digital Business & Enabling Technologies, https://doi.org/10.1007/978-3-030-39863-7_3

behaviour. This chapter outlines the RECAP approach to application optimisation and presents its framework for joint modelling of applications, workloads, and the propagation of workloads in applications and networks. The interaction of the models and algorithms developed is described and presented along with the tools that build on them. Contributions in modelling, characterisation, and autoscaling of applications, as well as prediction and generation of workloads, are presented and discussed in the context of optimisation of distributed cloud applications operating in complex heterogeneous resource environments.

Keywords Resource provisioning • Workload modelling • Workload prediction • Workload propagation modelling • Application optimisation • Autoscaling • Distributed cloud

3.1 Introduction

Key to the RECAP approach for application optimisation is *application autoscaling*, the dynamic adjustment of the amount and type of resource capacity allocated to software components at run-time (Le Duc and Östberg 2018). In principle, this type of scaling can be done *reactively*— by dynamically adjusting the amount of capacity to match observed changes in load patterns, or *proactively*—by operating on predicted future load values. Naturally, proactive autoscaling requires the ability to predict or forecast future values of the workloads of applications, systems, and components.

In this chapter, we summarise the RECAP application optimisation system. Following the problem formulation, we discuss the RECAP approach to application modelling, workload modelling, and the models used for application optimisation (application and workload, including how models are constructed and trained), the optimisation approach, and the implementation and evaluation of the optimisation models. The application optimisation approach outlined in this chapter exploits the

J. Domaschka
Institute of Information Resource Management, Ulm University, Ulm, Germany
e-mail: joerg.domaschka@uni-ulm.de

advanced techniques for characterising, predicting, and classifying workloads presented in Chap. 2 to construct proactive autoscaling systems.

3.2 Problem Formulation

The problem of optimising the deployment and configuration of applications hosted in geo-distributed resource environments can conceptually be viewed as a graph-to-graph mapping problem. As discussed in previous chapters, RECAP models distribute applications as graphs of components where graph nodes denote application components and the edges of the graph represent the communication paths and dependencies among components. Similarly, infrastructure systems can also be represented as graphs where the nodes correspond to resource sites and the edges model the interconnecting network links of site-connecting networks. The optimisation problem then is to find the optimal mapping of application nodes to infrastructure nodes. This mapping is subject to constraints that reflect requirements on the application level (e.g. minimal acceptable Quality of Service (QoS) for applications, or co-hosting restrictions of I/O-intensive processes).

A graph-based formulation of the mapping problem facilitates reasoning on the scaling of both application and infrastructure systems. Application scaling can on the one hand regulate the optimal number of instances to deploy for specific component-associated services (horizontal autoscaling) and on the other hand define how much resource capacity to allocate to a particular application on a specific site (vertical autoscaling). Furthermore, application scaling can be global, when the entire application is scaled, or local, when only individual components are scaled independently. Hybrid approaches are also possible where individual parts of applications or infrastructures are treated differently. In that respect, the RECAP optimisation approach includes the concept of application and infrastructure resource zones—subsets of application and infrastructure graphs that need to be treated as a group.

Based on studies of the technical trade-offs that influence optimality in scaling and placement, e.g, power-performance trade-offs and sensors and actuators that can used in optimisation of systems (Krzywda et al. 2018), we define four types of constraints on application and infrastructure placement and scaling:

1. Affinity constraints—these specify co-hosting or pinning of components;
2. Anti-affinity constraints—these prohibit co-hosting or pinning;
3. Minimal number of instance constraints—these specify lower bounds of the number of instances to scale or the amount of capacity to allocate to application nodes; and,
4. Maximal number of instance constraints—these specify upper bounds of the number of instances to scale or the amount of capacity to allocate to application nodes.

3.3   Optimisation Framework

As well as infrastructure optimisation, RECAP provides an optimisation framework that enables the development and execution of optimisation tasks at application level. The core of the framework is an optimisation engine that consists of multiple modellers and optimisers. More specifically, the modellers produce dedicated models for each supported application, including workload models, load transition models, user models, and application models. These are used to provide a complete view of the application. In addition, they are used as input for optimisers to solve optimisation problems related to autoscaling. Depending on the type of optimiser, it can deal with a wide range of optimisation problems related to the placement, deployment, autoscaling, and remediation of applications.

For creating the respective modellers and optimisers, RECAP uses methodological framework that entails three optimisation levels for the deployment and management of applications in heterogeneous edge-cloud environments, see Fig. 3.1. The figure illustrates the three-level process that constitutes the optimisation methodology: the first level of optimisation is the simplest and aims at the placement of applications throughout the edge-cloud environments under fixed network, application, and quality-of-service requirements/constraints. Optimisation solutions created by this level of optimisation can be used for long-term resource planning as well as initialisations for further optimisation levels.

In the next level of optimisation, the variations of workload and user behaviours are taken into account for dynamic application placement and autoscaling. The workload model and user models are used to estimate the demand of resources of individual application components over time.

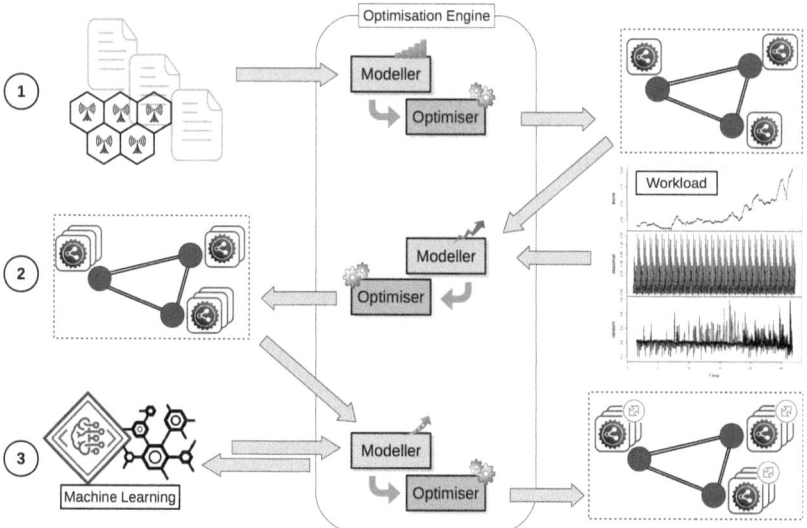

Fig. 3.1 A stratified approach to application optimisation iteratively building on three optimisation building blocks—(1) classic optimisation on static data, (2) application adaptation to variations in workloads and resource availability, (3) joint autoscaling and optimisation in multi-tenancy scenarios using machine learning (adapted from Le Duc et al. (2019))

With the estimation results, resources are allocated for each application component. Furthermore, based on predictions, workload can be redirected or migrated in order to maintain the load balance within applications.

The most advanced level of optimisation aims at proactive resource provisioning for applications. For that purpose, the RECAP Application Optimiser applies workload predictors that make use of workload models discussed in previous chapters. Machine learning is adopted to improve the understanding of both workload and application behaviours. This means more fine-grained models are derived and models can be refined and improved over time. Using these models, predictions can be performed more accurately to support load balancing, autoscaling, and remediation in a proactive manner.

3.4   Application Modelling

This section addresses the application modelling from an optimisation perspective. It first introduces the basic requirements and fundamental assumptions behind the RECAP application optimisation framework. Then, Sect. 3.4.2 introduces the modelling framework.

3.4.1   Application Characteristics and Modelling Requirements

A key enabling technology of cloud and edge computing is virtualisation. Abstraction of physical resources through containers and virtual machines (VMs) enables consolidation of compute capacity and resources (e.g. processors, storage, and networks) into software-defined infrastructures (SDIs). Such abstraction provides means for automation, scaling, and optimisation of resource allocations and resilience to variations in workload intensity.

The services that are running in an edge/cloud environment differ from the standard centralised cloud deployments. The difference is dictated in the way applications are used, by the infrastructure topology, and the infrastructure availability at the network edges. Geo-distributed infrastructures enable services (applications) to be brought closer to users, which increases the data exchange speeds and results in faster content delivery to consumers. However, the distributed nature of edge infrastructure comes with the limitation of physical space and associated limited hardware deployment capabilities. Different types of applications require different hardware profiles to process user requests. Varying hardware properties across distributed infrastructure stacks also require a distributed application architecture that is modular enough to adapt to the available edge/fog/cloud infrastructure horizontally and vertically.

As part of the application mode, the application topology depicts application components that can be deployed as separate entities (containers or VMs), and network link connections between them. A simple example of such a topology would be a deployment of an application that has a front-end web server and a database as two components. The web server can be deployed as a separate component on a separate VM or node, or even on a different datacentre than the database component, but both of them should have a bidirectional data flow connection for data exchange.

Most web applications serve different types of user requests, and to do so, different amounts of resources are needed depending on the requests made. For example, a request of streaming video from a content distribution network (CDN) would differ from a request to upload video to the service. The first request can assume a data download; however, the second request requires data upload, resampling, encoding, and other types of content analysis and optimisations.

Elasticity is one of the major benefits of virtualised resources. Elasticity makes it possible to maintain application QoS by dynamically scaling application components based on workload intensity (provided that the application architecture caters for it). To take advantage of elastic properties, application components can be managed by a load balancer that can spawn extra parallel application components and redirect user requests to evenly distribute the load.

Another vital characteristic in a distributed system is the geolocation and content variability within the same type of application. For example, a database or a CDN can be distributed across the edge infrastructure, where some instances will contain the same type of data, but some will not. Depending on data needed to serve a user request, the request needs to be routed to an application instance that has these data. Such scenario requires additional intelligence in the workload orchestration. From a RECAP perspective, it also requires that the application model have a notion of data content available within the application component.

All the aforementioned characteristics should be captured and reflected in the constructed application models. Specifically, the models should provide the means to estimate computation, memory, and storage capacity requirements of each components, as well as to present and calculate the mapping of the applications and application components on the underlying infrastructure. Moreover, they should help identify the type of traffic or content delivered to the users at different locations, and estimate the service delay for user requests.

3.4.2   Application Modelling Framework

The characteristics of the edge-cloud infrastructure and applications result in high complexity when it comes to application modelling. In particular, this modelling always has to be done in an application-specific manner. As such, it is necessary to have a comprehensive understanding of typical systems, models, and modelling tools from theoretical and practical perspective. This section outlines the strategy adopted in the RECAP methodology to perform application modelling for each system in its scope.

Firstly, a literature survey generated a universe of general and common architectures of distributed applications including client-server architecture, cloudlet, service-oriented architecture, and micro-services. Secondly, desk research allowed a comparison of the previously identified architectures with those of real large-scale distributed systems/applications and related industrial technology standards used for the realisation of operational systems. These systems include, for instance, the multi-tier caching system of Akamai's CDN, the operational architecture of Nextflix's CDN, the architectural framework of ETSI NFV, and Service Function Chaining (SFC) (Halpern and Pignataro 2015) amongst others.

Thirdly, following the literature survey and desk research, we explored mathematical models and tools widely adopted in computing science, such as queuing theory, graph theory, and control theory. These have been used to model components and the interaction among components and are used together to model applications. For instance, a model may be a combination of queuing theory, graph theory, and control theory as follows:

- Queuing theory is used to model the processing logic of a component, e.g. a VNF, a vCache, or a function node (database, data aggregator, balancer).
- Graph theory is used to model the network topology, service function chain, communications between components.
- Control theory is used to model the control logic of dispatchers, balancers, or orchestrators.

Fourthly, we decomposed the application into isolated components and analysed the components in detail in order to understand the nature of each component as well as how they communicate to each other in the whole application. In addition to simplifying the modelling task, this helps to identify how components impact each other and to identify the bottlenecks within the entire topology. Once all components have been modelled, these sub-models can be integrated to form a complete application model.

In order to keep the modelling effort manageable, it is further necessary to identify in advance the factors or metrics that should be captured in the models in accordance with the requirements of the application. That is, an application model always needs to capture business-specific constraints and goals and cannot be constructed on a technical level alone.

3.5   Workload Modelling

Besides an application model, an application optimisation engine further requires predictions of the amount of work the application shall be able to process overall or per zone. This enables proactive optimisation approaches, where readjustments do not happen on a best-effort basis upon changing workload conditions, but rather anticipate future workload levels and scale the applications accordingly.

As discussed in previous chapters, workload analysis and modelling focuses on techniques efficiently applied to time series data collected from both real production systems and emulating systems. Before being analysed, original time-series data have gone through a pre-processing step (gap filling, smoothing, resampling, etc.). Next, the data flows through the two-stage process of workload analysis and workload modelling.

Workload analysis is composed of two main tasks: workload decomposition (which splits a time-series into the trend, the seasonality, and the random factors of the workload data) and workload characterisation (which aims to extract workload features as an input and driver of workload models). Such workload characterisation is performed by an Exploratory Data Analysis (EDA), which is typically a manual step.

With this understanding of key features (metrics) of the workload, it is possible to derive which aspects should be considered in modelling, and how to effectively construct and evaluate appropriate workload models. To perform the modelling task, we adopt different categories of techniques in the RECAP methodology:

- Autoregressive integrated moving average (ARIMA) and seasonal ARIMA (SARIMA) models are chosen based on the analysis of the autocorrelation functions, partial autocorrelation functions, and tests of stationarity (for example, Dickey-Fuller tests).
- The family of autoregressive conditional heteroskedasticity models (ARCH, GARCH, NGARCH, etc.) is suited when the assumption is fulfilled that the variance of the time series is not constant but still a function of previous variances.
- Recursive neural networks and deep neural networks can be used to find more complex interactions between past and future requests.
- Long short-term memory neural networks (LSTM) shall be used when the aim is to detect if long past requests have a predictive value for future requests.

The models obtained by workload analyses are expected to provide forecasting capability. That is, based on them, it shall be possible to generate future workload predictions at different periods of time and different intervals, as required by the application under analysis.

The predictions can be validated through various metrics so as to identify the best one for each case, i.e. the one with the smallest errors. Depending on the application, different models are required for different prediction purposes (online and offline predictions). More specifically, models used for offline prediction are expected to provide high accuracy so that the results can be used for long-term planning.

In contrast, models used for online prediction additionally focus on low execution time besides the accuracy in order to offer short-term predictions in a timely manner. For the latter, it is necessary to evaluate the execution time in forecasting of each model and to balance between the error level and the execution time. Once selected, a model becomes the core of the online predictor component for that application within RECAP.

3.6   MODEL-BASED APPLICATION OPTIMISATION

The key challenges of efficiently deploying distributed applications in edge and fog computing environments involve determining the optimal locations and allocations of resource capacity. This is complicated by the inherently varying load conditions of distributed infrastructures. Due to the complexity and distinct mathematical formulations of the problems for different applications, they are typically treated separately, often with the solution of one as input to the other.

3.6.1   *Application Autoscaling*

In this section, we focus on application load balancing and distribution, which can be seen as a special case of holistic application autoscaling (i.e. self-scaling of application capacity allocations) given specific workload arrival patterns and component placements. To avoid confusion, we use different terms for different types of load balancing in distributed, multi-tier applications. We denote (1) load balancing between multiple instances of one single application component as load *balancing*, and (2) load balancing within the entire application to balance the loads distributed to different application components as load *distribution*. This is addressed by load transition models that capture how workloads flow through applications in workload models. Principally, RECAP assumes that a RECAP-enabled application is designed such that it is capable of making use of more instances (scale out) or more resources (scale up).

The RECAP application models describe applications as networks/ graphs of components with interdependencies and constraints in the form of network links, quality-of-service requirements, and communication patterns. Application components are split into front-end and back-end layers (modelling load balancing within components and management of component functionality respectively) that can be autoscaled independently. Using RECAP application and transition models, application workload arrival patterns can be used to derive how load propagates through distributed applications, and how the resulting component workloads impact the resources (including networks) where component services are deployed. Using prediction algorithms allows to provide improved performance and proactivity in autoscaling without changing the autoscaling algorithms themselves.

Depending on the application, arrival patterns can be measured from instrumented applications or infrastructure, or derived from simulation models build on the user (including mobility) models. The value of such models lies in the increased understanding of user and system behaviour, but also in their potential use for prediction of workload fluctuations in predictive scaling algorithms. Overall, RECAP applies the following types of autoscaling algorithms:

- Local reactive scaling algorithms (similar to the autoscaling algorithms used in Kubernetes) are used to individually scale component front-ends and back-ends. They apply varied degrees of downscaling inertia in order to reduce the amount of false positives.
- Global reactive scaling algorithms that predictively evaluate the performance of individual component autoscalers, and selectively apply those that maximise application objective functions, and used to control back-ends.
- Global proactive algorithms that use short time-frame simulation techniques to evaluate application performance for heuristically selected subsets for autoscaling actions. This class of algorithms shows the greatest autoscaling performance but is also significantly slower and resource demanding. This limits its applicability in large-scale systems.

3.6.2   Migration Techniques and Infrastructure Planning and Provisioning

In addition to the algorithms described above, which target application autoscaling in static deployment scenarios i.e. for deployments that do not change during or from autoscaling, RECAP also provides tools and techniques for evaluating and performing migrations of service components at run-time. In order to incorporate this functionality in the autoscaling-oriented perspective of application optimisation, such migrations can be formulated to be part of the autoscaling problem by scaling the amount of service instances for a specific component at a specific location in the space of $[0; n]$ (when there is no mobility of services and n is a maximum instance count). Alternatively, it can be treated as a separate placement problem that does not include autoscaling (beyond considering autoscaling limits in the placement process) that is solved independently (either before or after the autoscaling).

RECAP explores and exploits both approaches to develop a flexible framework for application migration optimisation. The basic building blocks of this framework are application autoscaling algorithms, a set of heuristic functions that evaluate alternative deployment scenarios (under specified autoscaling settings), and an in-situ simulation framework that models application communication at message level within the application models. The simulation framework essentially uses application and (predicted) workload models to simulate how workloads will be processed under current application deployment and autoscaling settings. Heuristic functions are used to identify components and/or resource sites that underperform or for some other reason are candidates for reconfiguration, and alternative configuration settings are speculatively evaluated by the simulator to identify the reconfiguration actions most likely to improve application performance the most (according to application heuristics and KPIs).

Currently three types of heuristic functions are used for deployment cost evaluations:

- Local evaluation functions that build linear combinations of QoS or KPI metrics for individual component services or resources,
- Aggregate local evaluation functions that define statistical aggregates of local evaluation functions for sets of components (i.e. subsets of application components) or resources (e.g. regions of resource sites), and

- Global evaluation functions that operate on application and infrastructure models to aggregate QoS or KPI evaluation functions for all components of an application or large sets of resources.

Evaluation functions are defined as mathematical constructs and can be composed to develop utility functions that combine evaluation of both applications and infrastructure resources. Using the simulation techniques, recommendations for how to change application deployments (in single- and multi-tenancy scenarios) and autoscaling constraints can be derived from nominal size estimations of component placements and infrastructure capacities, or conversely component nominal sizes can be included in the decisions on the admission of scaled or migrated service instances from autoscaling constraints.

3.6.3   Workload Propagation Model

Workload propagation models describe how workload is propagating through an application (i.e. between application components), and how a fluctuation of workload at a certain component impacts the other ones. Such a model can be constructed using workload data collected from all the network nodes/locations in every system. Unfortunately, due to the size and complexity of large-scale systems, exhaustively collecting such data is extremely challenging (Le Duc et al. 2019). Therefore, the mechanisms for workload generation and/or propagation are needed that enable the production of workload data for all network nodes using data traces collected only from a subset of nodes.

The five workload diffusion algorithms to address this problem are classified into non-hierarchical and hierarchical diffusion as follows:

1. Non-hierarchical diffusion—these algorithms perform load propagation within networks according to a discrete spatial model of how heat is diffused in materials in physics or chemistry. They are applicable for controlling data exchange and the workload of synchronisation tasks that are carried out by neighbouring network nodes. This also can be extended to cover some general cases of unstructured peer-to-peer overlays or ad-hoc mobile networks.

2. Hierarchical diffusion—these algorithms rely on a hierarchical network model which slices the network into different layers. In this case, the workload propagation is directed through network layers from end users down to the core network layers or using predetermined routing paths destined to the dedicated service nodes. This diffusion technique is applicable for core broadband networks and CDNs.

3.6.4   *Approach and Realisation*

Non-hierarchical diffusion algorithms include population-based, location-based, and bandwidth-based algorithms, while hierarchical diffusion ones include hierarchy-based and network-routing-based algorithms. This section briefly describes the five algorithms including the assumptions, key inputs, the main flow, and properties (see Table 3.1). Further details about the calculations or formulae used in each task/step of the algorithms can be found in RECAP Deliverable 6.2 which can be downloaded at the RECAP website. The algorithms were tested with workload data traces collected as time series at three inner-core nodes of BT's CDN; the representative metric of the workload in this use case is the traffic generated at caches when serving user requests.

3.7   The RECAP Application Optimisation Platform

Elasticity is a key function for addressing the problem of reliable resource provisioning for edge-cloud applications as it ensures the reliability and robustness of the applications regardless of the non-linear fluctuation of the workload over time (Östberg et al. 2017; Le Duc et al. 2019). One of the key techniques adopted in RECAP to address elasticity and remediation is autoscaling. By flexibly adjusting the amount of resources allocated for applications and/or the number of application instances or components, autoscaling enables applications to adapt to workload fluctuations, which helps prevent the applications from becoming unresponsive or terminating.

Table 3.1 Summary of diffusion algorithms

Diffusion algorithm	Assumptions	Key inputs	Description
Population-based Location-based Bandwidth-based	• Non-hierarchical network/application topologies (e.g. telecom networks, P2P applications) • Homo-geneous user behaviour	User distribution in the network Geographical node locations Bandwidth capacity of network links	• Iterative refinement algorithms (similar to heat diffusion and spring relaxation equations) • Repeatedly solve state equations to distribute workload to neighbours until the overall load distribution approaches equilibrium • Algorithms highly parallelisable
Hierarchy-based	• Hierarchical network/application topologies (e.g. broadband networks, CDN application) • Full mesh network of the inner-core nodes • Multiple shortest path routing	• Network hierarchy • Bandwidth capacity of network links • User distribution in the network	• Hierarchy-based user aggregation to identify the aggregated number of users at every node/location based on bandwidth capacity of neighbouring links • Backward workload extrapolation to collect the workload measurements from every node to the inner-code nodes • Inner-core workload extrapolation to extrapolate workload at every inner-core node (if needed) • Workload propagation to distribute the workload from inner-code nodes to every node in the network
Network-routing-based	• Homo-geneous user behaviour	• Network hierarchy • Bandwidth capacity of network links • User distribution in the network • A set of service (inner-core) nodes	• Routing path discovery to identify (shortest) routing paths from client-clusters to the service nodes • Network-routing-based user aggregation (using routing paths) • Backward workload extrapolation • Workload propagation

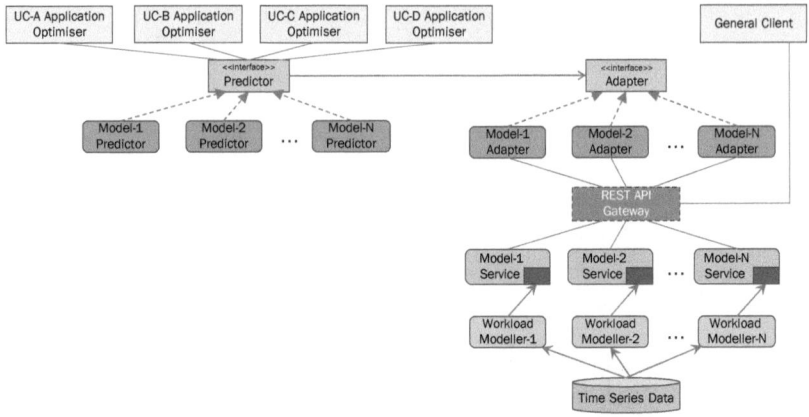

Fig. 3.2   A platform for the integration of predictors and modellers

The RECAP system model assumes that applications are dynamically distributed, and their behaviours are considerably difficult to predict and model. Moreover, each application component may be subject to different workloads, and an understanding of workload characteristics is required in order to autoscale efficiently. Therefore, workload analysis, modelling, and prediction based on time series analysis become the vital factor for the efficiency of our solutions, and especially so for the proactive schemes.

For application optimisation, RECAP has created a platform for the integration of application optimisers, workload modellers, and workload predictors, as shown in Fig. 3.2. The figure also shows how the optimisers (autoscalers) utilise the predictors when needed. Predictors, in turn, are fed by workload modellers.

By adopting different techniques, for example regression or machine learning, a workload modeller can construct multiple workload models. For flexibility in system integration, the modellers can be implemented using various technologies. The workload models are then wrapped and exported as a microservice using a REST API Gateway. On top of model services, a set of adapters are built to provide a unification layer. While workload models are constructed using historical workload data, they can be updated continuously at run-time by the workload modeller. Predictors in the platform make use of the adapters in order to access the available models and to make their predictions.

On top of this platform, robust and efficient approaches for autoscaling are constructed based on the results of workload modelling and prediction. Optimisers encapsulating optimisation algorithms and application-specific constraints make use of prediction for proactive optimisation. Note that an optimiser can call multiple predictors that access different models constructed using different techniques. This also implies that predictors are developed for every constructed model. Such implementation enables the capability of extension when a new optimiser (of a new application) or new model (using new techniques) is added to the system.

3.8   Conclusion

This chapter introduced the RECAP Application Optimisation approach and framework and outlined its the constituent building blocks. The interaction of the RECAP models and algorithms developed was further discussed. The RECAP Application Optimisation Framework addresses application placement and autoscaling, and provides models and tools for prediction, optimisation, and evaluation of the performance of distributed cloud applications deployed in heterogeneous resource environments.

References

Le Duc, Thang, and Per-Olov Östberg. 2018. *Application, Workload, and Infrastructure Models for Virtualized Content Delivery Networks Deployed in Edge Computing Environments.* Proceedings of the IEEE International Conference on Computer Communication and Networks (ICCCN), 1–7.

Le Duc, Thang, Rafael García Leiva, Paolo Casari, and Per-Olov Östberg. 2019. Machine Learning Methods for Reliable Resource Provisioning in Edge-Cloud Computing: A Survey. *ACM Computing Surveys (ACM)* 52 (5): 39. https://doi.org/10.1145/3341145.

Halpern, J., and C. Pignataro. 2015. *Service Function Chaining (SFC) Architecture.* RFC 7665.

Krzywda, Jakub, Ahmed Ali-Eldin, Trevor E. Carlson, Per-Olov Östberg, and Erik Elmroth. 2018. Power-Performance Tradeoffs in Data Center Servers: DVFS, CPU Pinning, Horizontal, and Vertical Scaling. *Future Generation Computer Systems* 81: 114–128.

Östberg, Per-Olov, James Byrne, Paolo Casari, Philip Eardley, Antonio Fernández Anta, Johan Forsman, et al. 2017. *Reliable Capacity Provisioning for Distributed Cloud/Edge/Fog Computing Applications.* European Conference on Networks and Communications (EuCNC).

CHAPTER 4

Application Placement and Infrastructure Optimisation

Radhika Loomba and Keith A. Ellis

Abstract This chapter introduces the RECAP Infrastructure Optimiser tasked with optimal application placement and infrastructure optimisation. The chapter details the methodology, models, and algorithmic approach taken to augment the RECAP Application Optimiser output in producing a more holistic optimisation, cognisant of both application and infrastructure provider interests.

Keywords Resource provisioning • Application placement • Infrastructure optimisation • Infrastructure modelling • Distributed clouds

R. Loomba • K. A. Ellis (✉)
Intel Labs Europe, Dublin, Ireland
e-mail: r.l.loomba@ieee.org; keith.ellis5@mail.dcu.ie

69
T. Lynn et al. (eds.), *Managing Distributed Cloud Applications and Infrastructure*, Palgrave Studies in Digital Business & Enabling Technologies, https://doi.org/10.1007/978-3-030-39863-7_4

4.1   Introduction

As discussed in Chap. 3, the RECAP Application Optimiser derives initial placements from network topologies. These placements utilise application, workload, and workload prediction models to derive scaling models, which are combined with Machine Learning (ML) tools to produce application models and recommendations. So, one might think "given I have an application placement recommendation, what does the application placement and infrastructure optimiser do?"

Well, put simply, the application optimiser does not have all the requisite information to make an optimal decision. When considering an application placement decision, one must take into account that typical operations perceive a separation of concerns between an application provider and an infrastructure provider, the latter typically dealing with multiple application provider requests.

In essence, the RECAP Infrastructure Optimiser brings the interests and heuristics of the infrastructure provider to bear on the overall placement decision and is tasked with three functions:

1. Application Placement: the mapping of individual application service components to individual infrastructure resources. The focus being to identify, traverse, and optimally select from possible placements.
2. Infrastructure Optimisation: focuses on optimising the availability and distribution of an optimal number, type, and configuration of physical resources, while optimising their utilisation, i.e. "sweating the assets".
3. Capacity Planning: considers future workloads and decides what type of physical resource/node should be placed in the network, how many nodes, and where to place them.

The remainder of this chapter is structured as follows. Section 4.2 briefly outlines the RECAP Infrastructure Optimiser architecture followed by the problem formulation for making a more holistic placement optimisation decision. We then present and discuss three different models used in the RECAP Infrastructure Optimiser, namely load distribution models, infrastructure contextualisation models, and load translation models. Finally, we outline the RECAP algorithmic approach for optimal placement selection.

4.2   HIGH-LEVEL ARCHITECTURE
OF THE INFRASTRUCTURE OPTIMISER

Functionally, the RECAP Infrastructure Optimiser as presented in Loomba et al. (2019), can be considered in terms of an "offline infrastructure optimisation modelling" process, Fig. 4.1, and an "online infrastructure optimiser" implementation, Fig. 4.2.

The offline infrastructure optimisation modelling process and the online implementation components/steps are illustrated within the grey

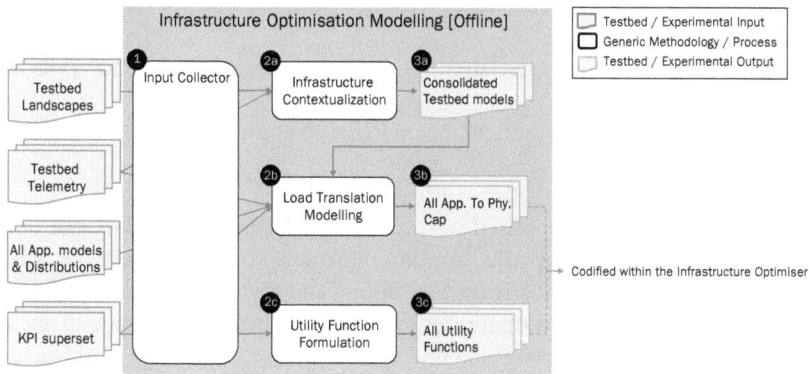

Fig. 4.1   Offline infrastructure optimisation modelling process

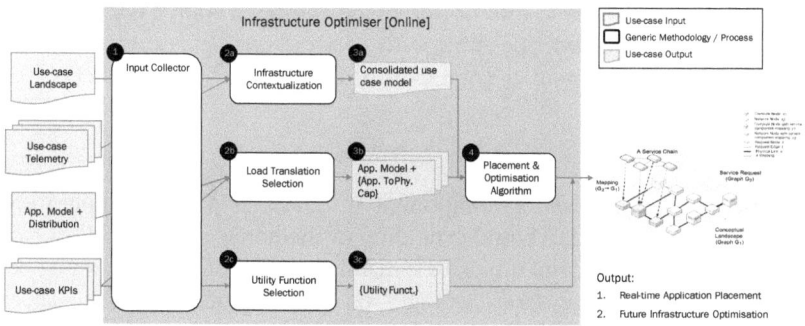

Fig. 4.2   Online application placement and infrastructure optimiser

boxes in Figs. 4.1 and 4.2, although it should be noted that infrastructure optimisation is highly dependent on the veracity of its inputs as depicted by the leftmost boxes in Figs. 4.1 and 4.2.

The main steps in the RECAP infrastructure optimisation process are outlined in Table 4.1 below.

4.3   Problem Formulation

The application placement and infrastructure optimisation challenge is threefold, i.e. how to:

- Optimally match the requirements of components, e.g. VMs to physical resources
- Adhere to SLAs negotiated between the infrastructure and application providers
- And when to instantiate the virtual components relative to the required capacity

Each challenge is composed of several subchallenges, including but not limited to determining the optimal abstraction of components and resources, defining the objectives that bound the placement, and identifying a common means to compare deployment solutions. This may, in fact, involve multiple-competing objectives; hence, there must be a trade-off.

Current literature suggests stochastic bin-packing method (Jin et al. 2012), approximation algorithms (Meng et al. 2010), and a variety of other heuristics (Li et al. 2014) with focus on specific resources or objectives, e.g. resource usage, and power and thermal dissipation (Xu and Fortes 2010). However, often commercial and open source orchestration solutions schedule either pessimistically to avoid conflicts or opportunistically to gain from potential Total Cost of Ownership (TCO) benefits.

4.3.1   Infrastructure

Deploying applications or component instances as VMs or application containers, requires a rich understanding of the heterogeneity and state of the underlying infrastructure. This is mainly because the application workloads might be computation-, storage-, or network intensive. With respect to infrastructure, the requisite information is represented as a multi-layered graph of the physical and logical topology called a "landscape"

Table 4.1 Steps in the RECAP infrastructure optimisation process

Step No.	Step title	Description
1	Input collection	Manages the data ingress of inputs and is essentially the same for offline modelling and the online implementation; what changes is the context. Inputs include: • Composition and structure of infrastructure available, i.e. landscape • Composition and structure of the application (e.g. application model and load distributions) • Associated telemetry data, e.g. from testbed or system of interest • Infrastructure provider and service/application provider KPIs
2	Modelling and/or selection	Creates/utilises models primarily in the offline mode to produce outputs that are subsequently codified within the online optimiser. (2a) Combines telemetry with an infrastructure landscape and filters as appropriate based on relevant KPIs. The process is the same for offline and online, only the context changes. Offline (2b) creates application load translation models, which map how application load correlates to resources, associated telemetry and/or KPIs. Online this step more simply involves appropriate selection of a subset of "application load to physical capacity mappings" from those modelled offline. Offline (2c) utilises KPIs and Multi-Attribute Utility Theory (MAUT) to formulate Utility Functions for application and infrastructure providers. Online this step selects a subset of utility functions from those previously formulated offline.
3	Modelling/ selection output	Offline (3a) represents a consolidated infrastructure model, i.e. a testbed specific landscape that feeds the "load translation modelling" process. Online this is the use-case-specific landscape and is fed directly to the algorithm module step 4. Offline (3b) encompasses the complete set of possible "application load to physical capacity mappings" based on the testbed inputs. Online it is an appropriate subset of those modelled offline based on the use case inputs. Offline (3c) is the complete set of possible "Utility Functions". Online it is an appropriate subset based on the given use case.
4	Algorithmic optimisation	This step is illustrated in Fig. 4.2 and takes (3a) and (3b) as inputs. The algorithm subsequently provides several valid solutions, over which the utility functions selected in step (3c) are applied to select a near-optimal option. The output of step 4 is a real-time application placement or a future infrastructure optimisation.

(Metsch et al. 2015). This landscape is built utilising the Neo4J database. It is primarily a graph describing a computing infrastructure, that also details what software stacks are running on what virtual infrastructure, and what virtual infrastructure is running on what physical infrastructure. The data within a landscape is collected via collectors and event listeners. Collectors are provided for physical hardware information (via the hwloc and cpuinfo files) and for OpenStack environments.

This rich representation helps to understand the capability of the infrastructure. It is mathematically quantified as a landscape graph $G = (\ N^*, E^*)$ where N^* is the superset of all nodes of all geographical sites indicating resources such as CPU cores, processing units, and switches and E^* is the superset of all links between them, which might be communication links or links showing a relationship between two nodes.

Although this granularity of information is required, it increases the complexity of problem in terms of possible placement combinations and adds additional dimensions. For example, instead of determining at the aggregate server level, one must determine the cores, the processor bus, and the processing units involved in the mapping. As such, for simplicity, the landscape graph is abstracted in this initial formulation into a contextualised landscape graph $G_1 = (X, E)$ where $X \subset N^*$ and set $E \subset E^*$ containing only two categories of node, namely network and compute. Set X is a collection of nodes with a compute or network category and Set E is the set of all links connecting these nodes. This abstraction defines a network node to be of type, e.g. nic, vnic, switch, router or SRIOV channel. A compute node is defined as a resource aggregate of a given type, e.g. machine or virtual machine and is created by collapsing the properties of the individual compute, memory, and storage entities directly connected and contained within the node. This helps isolate the two categories of nodes while storing pertinent information regarding the other categories.

Building on work and experience of the Superfluidity project (Superfluidity Consortium 2017), these nodes also contain attributes which quantify their capacity. This is represented in a vector format as v_x for node x along with telemetry information regarding utilisation (the average percentage of times a resource is deemed to be "busy" over a predefined time window for the given resource), saturation (the average percentage of times a resource has been assigned more tasks than it can currently complete), cost models, etc. The superscripts c, m, n, and s denote compute, memory, network, and storage category values respectively. The compute node ς has capacity $v_x = \left[v_x^c, v_x^m, v_x^s \right]$ where v_x^c

represents the number of cores and v_x^m the amount of free or unused RAM from the total installed on the resource aggregate, and v_x^s represents disk size. A network node $\mathfrak{n}$ has a 2-tuple capacity vector $\mathbf{v}_x = \left[v_x^n, v^* \right]$ calculated based on its available bandwidth v^n and available connections v^*.

Furthermore, for physical communication link $e \in E$ representing the graph edge, link attributes are added including geographical distance $len(e)$ and measured or observed throughput τ_e and latency l_e normalised for time δt, just before any application placement decision is made. Values such as B_e for maximum bandwidth and associated rate R_e^b for a unit of bandwidth are also included in the infrastructure graph G_1.

4.3.2   Application

The application to be deployed is itself described as a service request, composed of either multiple connected service components in the form of service function chains (SFC), or disjointed service components, which need to be placed together. Using the definitions presented in (RECAP Consortium 2018), the application is represented in a service request graph $G_2 = (Z, F)$ with nodes represented by set Z and graph edges by set F. For this model, the nodes are further categorised similar to the method described above into either compute or network nodes and are termed as request nodes and request graph edges as they represent the service request.

4.3.3   Mapping Applications to Infrastructure

Each service component of the service request graph G_2 is then mapped on to infrastructure resources and links in graph G_1. This mapping is composed of a subset of nodes and graph edges of graph G_1, and as such it is important to first define the rules of such a mapping. This is represented in Fig. 4.3.

The nodes in graph G_2 are defined as a 1:1 mapping with resource nodes, defined by the set of infrastructure nodes Υ. This set $\Upsilon \subset X$ contains all compute and network nodes which have a service component mapped to them. This is also quantified as $z \to y$ where request node $z \in Z$ is mapped to resource node $y \in \Upsilon$, also ensuring that $Z \cong \Upsilon$. There is also a 1:N mapping for the graph edges as they get mapped to a set of

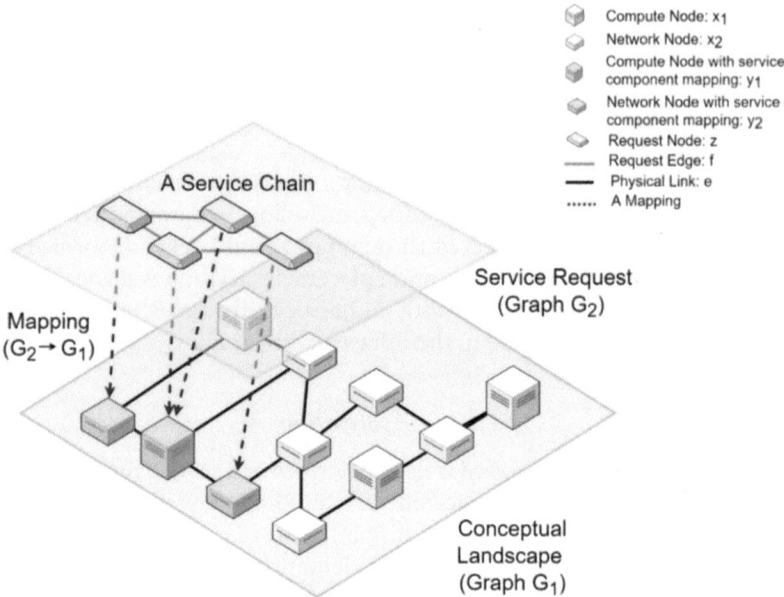

Fig. 4.3 Mapping a service request graph and a contextualised landscape graph

communication links also indicating a physical network path. This is defined as graph edge $f \in F$ being mapped to the set of communication links or path g, which is a subset of all possible paths in the infrastructure graph, also quantified as $f \rightarrow g$. These definitions describe a possible placement solution, denoted by $G_2 \rightarrow G_1$, which may or may not fulfil the criterion of an optimal mapping.

4.3.4   Mapping Constraints

The RECAP Infrastructure Optimiser's objective is to identify an optimal mapping, and this requires the analysis of several constraints. A subset of the most relevant constrains to be considered for an infrastructure resource request model along with its associated policies is presented here. These include the capacity requirement, compositional, SLA, and infrastructure-policies constraints.

4.3.4.1 Capacity Requirement Constraint

This constraint defines the capacity, specified in terms of compute, memory, network and/or storage capacity that must be available on the resources and edges that are intrinsically part of the mapping. For the initial formulation, this capacity is considered static during the duration of the application deployment.

For compute, this includes request information from the application for customised virtual compute flavours, related to the software image and the number of requested cores. Additionally, details on acceleration capabilities, CPU architecture, or clock frequency requests may also be included. For memory, this specifies the amount of virtual memory that needs to be allocated to the request node, and also includes details on whether Non-Uniform Memory Access (NUMA) support is required. For network, this includes request information on the required network bandwidth, the requested network interface (e.g. normal virtual NIC vs. SR-IOV) and additional acceleration capabilities. For storage, this includes request information on the required storage size, type of storage (e.g. SSD vs. HDD), and redundancy level (i.e. RAID).

In this scenario, it is imperative to remember that since the request edge is mapped to a set of communication links or a physical path, the aggregated bandwidth and aggregated latency of all edges that are a part of this physical path must meet the requirement. There are a number of reasons why this requirement must be met including propagation delay, serialisation, data protocols, routing and switching, and queuing and buffering. Of these, the most significant ones are the propagation delays and the queuing delays. Since the network devices are considered in the model, the saturation value of the nodes on the physical path are summed to get the queuing latency, and the aggregation of the bandwidth and propagation latency are quantified as follows.

- The aggregated bandwidth of the physical path is the minimum observed throughput of all edges in the physical path. Mathematically, $b_g = \min_{e \in g} \tau_e$
- The aggregated latency of the physical path is the summation of the propagation latency in the physical path. Mathematically, $l_g = \sum_{e \in g} l_e$

4.3.4.2 Compositional Constraint

Compositional constraint defines any rules that explicitly dictate the composition of the mapping at different levels of granularity such as resource types (e.g. compute and network), resource groups (e.g. a set of compute resources), or resource zones (e.g. a set of machines deployed at a particular location). At each level of granularity, the constraint can be further quantified as being affinity ("same" or "share") or anti-affinity specific ("not-same" or "not-share"). These dictates whether resources "share" the "same" physical resource, set of resources, properties or zones, or not.

The first example of such a constraint is the resource type mapping constraint; e.g. a virtual CPU (vCPU) must be mapped to a CPU or a virtual network interface (vNIC) must be mapped to a port of physical network interface (NIC). These are necessary since the network interface to which the vNIC is mapped needs to be on the same server as the CPU host, whereas depending on the configuration of the infrastructure and level of redundancy, the virtual storage component or disk can be mapped to a remote storage disk or to the local storage. Additionally, physical memory banks connected to the physical core allocated to the request deliver different performance in comparison to allocating memory banks connected to other physical cores.

4.3.4.3 Service Level Agreement Constraint

This constraint relates to specific customer service requirements and covers information regarding the scheduling prioritisation of individual service customers and application instances, as well as control policies to be enforced to pre-emptively suspend/kill other currently deployed services by the same or a different service customer.

This is modelled by defining a set of SLAs denoted by set S, negotiated between the infrastructure provider and the service provider/customer. The infrastructure provider creates and offers more than one SLA template. These are arranged in a hierarchical manner, with each level relating to a specific "type" of SLA such as "platinum", "gold" or "silver". Each template $s \in S$ is associated with one type that defines its rate R_s, the threshold of service level KPIs required by the customer for that template, the unit cost of SLA violations $j \in J_s$ for each KPI, and the list of failure-tolerant implementations $h \in H$ that need to be made by the resource provider.

The customer selects the setting they want within the chosen SLA template and can select more than one SLA for different applications. This becomes the agreed SLA for the application for the customer and is included as a customer request. As such, it also includes the total run-time (in hours) for the application instance T_r and other constraints that the customer requests.

4.3.4.4 Infrastructure-Policies Constraint

Apart from scheduling prioritisation, it is also important for the resource provider to define policies and control protocols for the management of the infrastructure. These constraints include resource allocation prioritisation and allocation ratios.

The resource provider can prioritise certain resource types or groups, based on either their cost or performance. These resource groups can also be associated with a particular SLA template and with allocation restricted in certain situations. Also, policies related to overprovisioning of resources need to be defined by the resource provider. This controls the ratio of allocating virtual resources to the physical resource and may differ by category of resource and specific use case. Additionally, it includes the minimum and maximum capacity that is allocated to one instance over the entire run-time of the application instance, if any.

4.4   MODELS THAT INFORM INFRASTRUCTURE OPTIMISATION DECISIONS

As discussed earlier, three models are used in the RECAP Optimisation Framework to inform optimisation decisions—load distribution models, infrastructure contextualisation models, and load translation models. As load distribution models have been discussed earlier in Chap. 3, we will focus on infrastructure contextualisation models and load translation models here.

4.4.1   Infrastructure contextualisation models

While "the map is seldom the territory", a good map invariably helps. As discussed earlier, the "infrastructure representation (landscape)" is an important input to the RECAP Infrastructure Optimiser and aims to provide a rich representation of the resource composition, configuration, and topology of the various entities in the cloud/edge infrastructure, across three layers—physical, virtual, and service. However, a landscape for a given scenario may not have all the requisite data expected (e.g. geographical and capacity), or it may be too rich having redundant information irrelevant to the specific use case. Additionally, it will not have telemetry data needed to support the optimisation process, e.g. current utilisation. Furthermore, if granularity of the infrastructural information is increased, multiple different mappings need to be considered, increasing the complexity of the NP-hard problem.

To address this issue, a "contextualised modelling process" (Figs. 4.1(2a)/4.2(2a)) is undertaken so as to produce a "consolidated infrastructure model" (Figs. 4.1(3a) and 4.2(3a)). This process may augment and/or filter the landscape input, adding telemetry, e.g. resource utilisation, KPIs (perhaps to filter), and important platform features identified for the individual use cases. But assuming the landscape provided meets the requirements of the optimiser, the only addition for creation of the consolidated model is telemetry, plus any required filtering based on appropriate KPIs. The type of information encapsulated in the consolidated model is illustrated by way of the following brief example. Figure 4.4 represents the network topology for the city of Umeå, Sweden.

The contextualised infrastructure model for such a network includes:

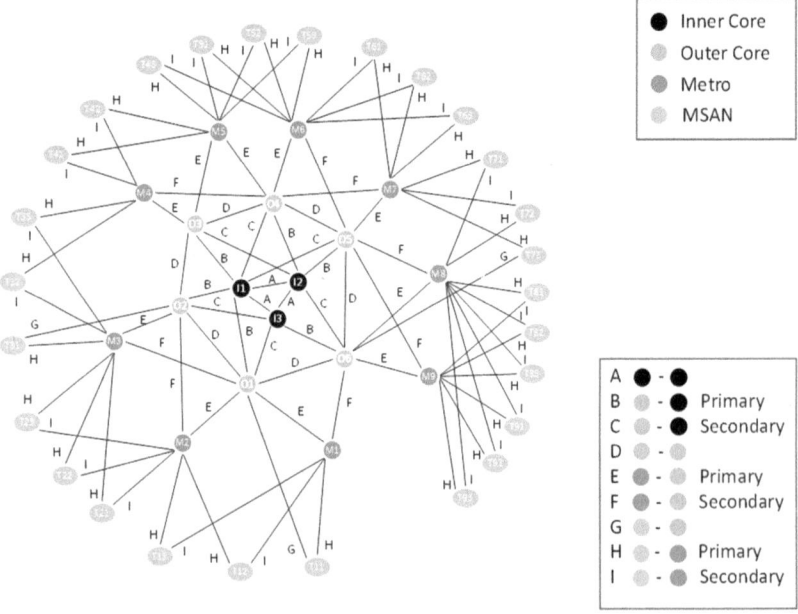

Fig. 4.4   Overview of Umeå network topology (site routers only)

- The definition of resource sites (e.g. MSAN and Metro), i.e. information pertaining to individual physical infrastructure elements, their physical attributes, and configurations, the communication links between them, and the properties inherent to these links.
- The definition of inter-site network bandwidth and latency, i.e. the available network bandwidth and latency values of the physical communication links between resource sites.

The output of this process is a representative graph of the network landscape. At a more granular level, Fig. 4.5 below shows the modelled communication links across the tiers for just one resource site T21, as presented by the Neo4J database.

4.4.2   Load Translation Models

Given a good understanding of the physical infrastructure, one must then consider the applications that are to be optimally deployed. In that regard and building on Chap. 3, load translation models serve to:

1. Quantify the association between virtual machine/container configurations and specific infrastructure configurations, and
2. Determine the lower and upper bounds on resource consumption in relation to varied application performance KPIs.

The RECAP load translation models are designed to be generally applicable to distributed application deployments and cloud/edge

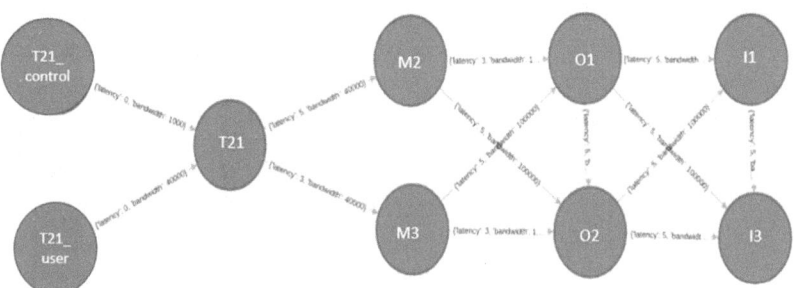

Fig. 4.5   Representation of a subgraph of contextualised network landscape

infrastructures. The methodology presented here focuses on mapping the virtual workload of individual service components to a set of prioritised time series data-points, i.e. telemetry collected from the physical infrastructure components. The correlation of service components and telemetry informs an offline modelling process that builds insights about application placement and related performance of deployed components. These insights which are then available to the online optimiser.

RECAP experimentation highlighted how understanding the profile mix of infrastructure features coupled with the different service components, e.g. Virtual Network Functions (VNFs) deployed on the infrastructure impacts the efficient usage and distribution of physical resources, i.e. compute, memory, network, and storage.

The RECAP load translation methodology is illustrated in Fig. 4.6.

The methodology begins with understanding the following inputs:

- **Infrastructure Information**: It is important to understand the testbed/experimentation set-up and to understand its limitations, especially those relating to the heterogeneity of available server configurations. This helps define the information on the infrastructure, e.g. physical machine, virtual machine and container configurations, total number of machines, and their connections.
- **Associated Telemetry Data**: A telemetry agent is initialised to collect data over multiple domains, e.g. compute, network, storage, and memory. All incoming data from these metrics are aggregated and normalised based on the domain context and studied as time series data.
- **Virtual Machine Configurations**: The application to be placed on these machines needs to be understood, how many instances and types of instances can be run, whether they are deployed as servicechain or disjoint VMs/containers, along with the various configurations of these VMs/containers. This information is typically provided via the application optimiser.
- **End-User Metadata**: The end-user behaviour to be emulated is determined based upon use case definitions and validation scenarios, and this specifies how users will access the applications, how many users will be initialised, how the number of users will increase/ decrease, and how different user behaviours will be simulated in the testbed.

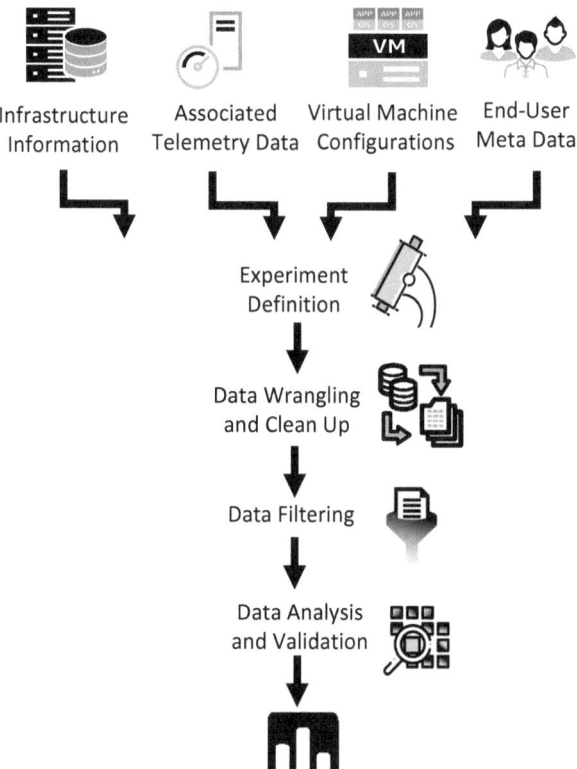

Fig. 4.6 The load translation methodology in full

These four inputs are then used to define the context of experiments that have to be run which are based on, for example, the duration of each experimental run; the prioritisation of configurations; application instances and end-user workloads that will be varied; and the number, type, and behaviour of users that will be applied for each experiment, as well as the number of times the same experimental set-up will be replicated for redundancy purposes. This helps define a set of profiles that are given to the load translation model to assess and analyse.

Once the experimental data is available and the experiment defined, the following data analysis steps are undertaken:

- **Data Wrangling**: The collected data is isolated and labelled appropriately according to experimentally relevant timestamps.
- **Data Filtering**: Files are filtered and integrated as appropriate for comparison.
- **Data Visualisation and Analysis**: Visualisation and analysis are completed for each defined experiment. The appropriate metrics are defined and calculated per machine/VM/container as well as device names, e.g. for network interfaces and storage devices attached to the physical machine. The results are then summarised and are visualised as a comparison between the average usage for compute, memory, network, and storage resources.

For illustration purposes, Fig. 4.7 shows compute utilisation for two machines (compute-1 and compute-3) for the different VNF placement profiles of Use-Case A, normalised given the number of cores of each machine. This type of analysis provides an initial and basic understanding of the relation between the workload, application, and infrastructure. But as more and more data gets collected and analysed, the accuracy of the normalisation across experiments, quantification of the relationship, and formulation of the mathematical equation for same increases. The results of this process are collated as a complete set of mappings to be used for specific use case optimisations.

Completing this may still seem relatively straightforward until one considers the different KPIs, constraints, complexity, and scale to be addressed.

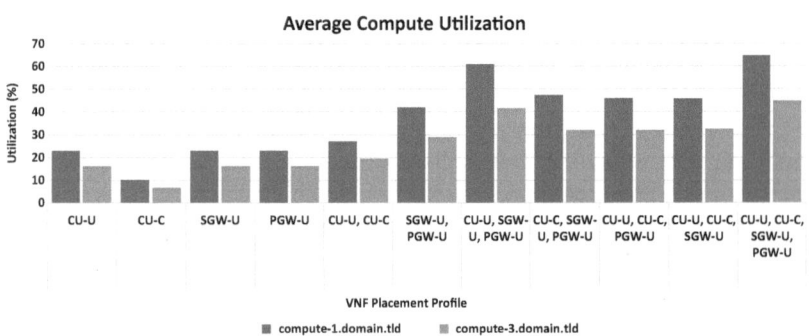

Fig. 4.7 Average compute utilisation by VNF placement profile for compute-1 compute-3

Given such scale and complexity, understanding, traversing, and promptly selecting from such nuanced options at scale must be mathematically derived and programmatically implemented, and this is the task of the algorithmic approach discussed next.

4.5   Algorithmic Approach to Optimal Selection

It should be apparent that application placement and infrastructure optimisation is highly dependent on the veracity of inputs received and that the optimiser is a collective of components and models coupled with the algorithmic approach applied to the output of those models, not just an algorithm. As such, this section describes (1) the utility functions, and (2) the evolutionary algorithm used in the RECAP Infrastructure Optimiser. The former is used as a uniform mathematical framework to normalise business objectives to compare possible placements identified by the algorithm. The latter was chosen for its appropriateness in quickly identifying and selecting near-optimal placement options.

4.5.1   Utility Functions

Previously the application placement problem was defined as optimally matching service requests to the capabilities of available resources, instantiating these components with the required capacity, and finally meeting SLAs between the resource and application providers. This transforms the problem into iteratively mapping individual service components on to various available infrastructural resources while meeting the constraints defined above. Solving this, presents many possible placement solutions out of which the optimal solution needs to be selected. Moreover, the number of possible solutions increases with growing sizes of application/ service requests and infrastructure, further increasing the complexity.

Determining the optimal solution is thus intrinsically challenging as it entails comparing deployments based on their benefits to either the provider or the service customer over a large solution space in both time and space. Furthermore, distinct yet complementary objectives and constraints must be handled, and trade-offs made. These objectives are often in different scales, ranges, and units, and need to be normalised into a common reference space.

Enter "Utility Functions", a key mechanism enabling analysis across varied objectives for different placement options and focused on

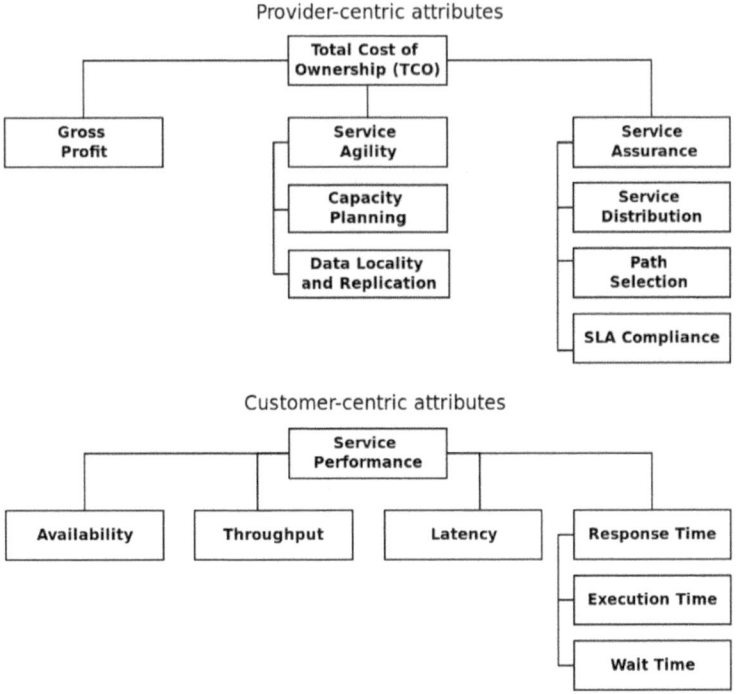

Fig. 4.8   Provider-centric vs customer-centric attributes (Loomba et al. 2017)

understanding the "reward" that is acquired per objective. Loomba et al. (2017) quantified these benefits as two utility functions, one for the resource provider denoted as U_P and one for the service customer denoted as U_c. These objectives are classified as being Provider-centric and Customer-centric, in Fig. 4.8.

1. **Provider-centric Objectives**: These objectives relate to the resource provider and deal with the management of the entire infrastructure. Key objectives include:
   - Gross Profit: This objective includes calculated revenue and expenditure costs for resource capacity allocated, cost of associated SLAs and SLA violations, and other costs of concern to the infrastructure provider.

- Service Distribution: This objective includes the analysis of available capacity of resources after the application deployment.
2. **Customer-centric Objectives**: These objectives relate to the service customer, quantifying Quality of Experience (QoE) as well as Quality-of-Service (QoS). These also help reason over the goodness of a placement decision and key objectives include:
   - Throughput: This objective includes the quantification of observed throughput over all physical links in the physical network, along with the analysis of dropped packets.
   - Latency: This objective includes the quantification of observed latency over all edges in the physical network, along with the analysis of packet delays.

This approach is extended in the RECAP methodology to consider enhanced constraints and use case definitions in defining a combined utility function that negotiates trade-offs between these two utility functions. This combined utility incorporated the preferences and priorities of the various use case business objectives and was evaluated using the Multi-Attribute Utility Theory (MAUT).

In this formulation, a multiplicative function is used to capture the interdependence of k conditional utilities for each attribute $a_i \in \mathcal{A}$. Here $\mathcal{A} = \mathcal{A}_1 \cup \mathcal{A}_2 = \{a_1 \ldots a_i \ldots a_K\}$, with $k \geq 2$ is the set of all objectives under consideration with subset $\mathcal{A}_1$ containing all provider-centric objectives and subset $\mathcal{A}_2$ containing all customer-centric objectives. For each of these conditional utilities, α_k indicates weight or a priority value of the objective while β_k is an additive weight that stores dependence on other objectives.

$$U_P = f\left(\alpha_i \cdot \mathcal{U}\left(a_i\right) + \beta_i\right), \text{where } a_i \in \mathcal{A}_1$$

$$U_C = f\left(\alpha_i \cdot \mathcal{U}\left(a_i\right) + \beta_i\right), \text{where } a_i \in \mathcal{A}_2$$

By assigning prioritisation weights to the provider utility and customer utility, the total utility of the placement can be calculated. These weights must be modifiable as they bias the selection of the placement solution. The total utility of the deployment with w_1 weight to the provider utility and w_2 weight to the customer utility can thus be defined as a weighted summation of the inputs. The optimisation function is then defined to maximise the total utility of the placement, considering minimum and maximum thresholds for the provider utility and customer utility. This is designed to facilitate graceful expansion to accommodate any variables/ parameters outside the scope of the current formulation or that gain importance to the use case owner following this analysis. The optimisation function is represented as follows:

$$\text{Maximise}: w_1 \cdot U_P + w_2 \cdot U_C$$

Subject to:

1.
$$\min_{\text{threshold}} (U_P) \leq U_P \leq \max_{\text{threshold}} (U_P)$$

2.
$$\min_{\text{threshold}} (U_C) \leq U_C \leq \max_{\text{threshold}} (U_C)$$

These use case specific utility functions thus ensure that the optimiser can adapt and reason over placement decisions even when business objective weights or priorities are changed.

4.5.2 Algorithms for Infrastructure Optimisation and Application Placement

While utility functions provide a mathematical framework for comparison, an initial challenge exists in being able to quickly select a subset of optimal solutions from the large number of possible deployment solutions for comparison.

In addressing this challenge, a *stochastic evolutionary algorithm* was selected. Its appropriateness relates to its incorporation of enough randomness and control to support decision making even for functions lacking continuity, derivatives, linearity, or other features.

Additionally, its ability to exploit historical information makes the algorithm much more efficient and powerful in comparison to exhaustive and random search techniques. Its advantages further include the ability to isolate a set of "good" solutions instead of just one, the possibility of parallelisation to improve efficiency and the support for multi-objective problems.

The algorithm calculates the optimal solution(s) in an admissible region for this combinatorically complex problem, which otherwise could not be solved by polynomial time. The optimality of the solution is based on its quality criterion called the "fitness function" and is represented as $f_{G_2 \to G_1}$ for the deployment solution $G_2 \to G_1$. This value is composed of the fitness of the individual mapping, based on the constraint definitions presented above. In the given application placement scenario, the objective of the algorithm is $\min f_{G_2 \to G_1}, \forall x \in [G_2 \to G_1]$ to ensure fast convergence for solutions that do not meet the required constraints. The output of the algorithm is thus the placement solution (or set of placement solutions) with minimum fitness or with fitness tending to zero (whichever is lower).

An overview of the algorithm is outlined in Fig. 4.9 and the accompanying text:

The set of possible placement solutions represents a "population", where each of these solutions is a "candidate" for the algorithm. The individual mappings in the placement solution are treated as "genes" (e.g. the mapping of a service component to an infrastructure resource).

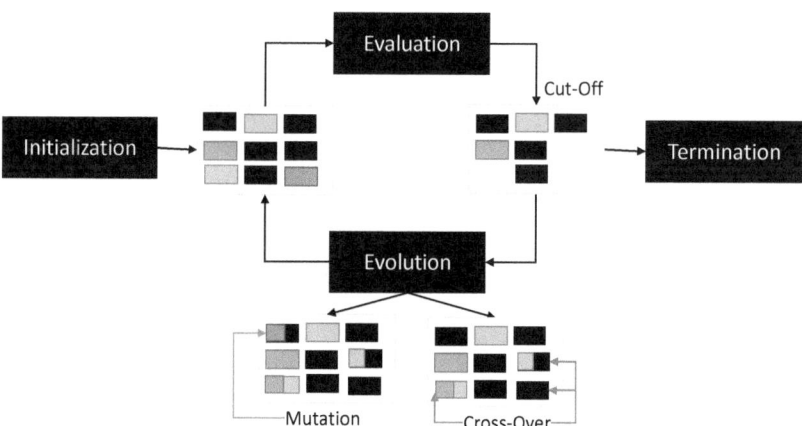

Fig. 4.9 Representation of the evolutionary algorithm

Initialisation: The algorithm is defined by a set of placement solutions or a set of candidates, which defines the current "population". These candidates can be arbitrarily chosen or can be based on prior knowledge from the use case to encompass a wide range of solutions. This is essential as with many different mappings from the placement solutions or "genes" present, it becomes easier to explore several different possibilities during the algorithm.

Evaluation: Each candidate in the population is evaluated according to the defined fitness function which numerically represents the viability of the solution. Thus, the next step is to eliminate the portion of the population with worse fitness values. The population now contains fitter "genes" or those mappings that fulfil defined placement criterion and have low fitness values.

Evolution: This step involves two main operations, mutation and cross-over. Both these operations are probabilistic and are used to create the next population from the remaining candidates.

- Mutation is used to introduce new candidates in the population by spontaneously changing one gene. In RECAP, this means that multiple mappings within a possible placement solution are swapped to create a new possible placement solution.
- Cross-over creates a mix of the available genes in the population by merging the genes from two candidates. In RECAP, this means that two mappings of different service components are taken from two different possible placement solutions and combined to create a new possible placement.

In either case, the resulting candidate may have better or worse fitness than existing members.

Termination: The algorithm ends in two cases. First after 30 iterations of the algorithm, for statistical accuracy and the second when it has found deployment solutions with fitness value equal to zero. All identified solutions are returned and evaluated and the solution with highest total utility is selected.

4.6   Conclusion

The RECAP Infrastructure Optimiser and the RECAP Application Optimiser are interdependent components in the RECAP ecosystem. The RECAP Application Optimiser does not have all the requisite information for an optimal decision, especially when considering that the typical mode of operation perceives a separation of concerns between the application and infrastructure providers. Therefore, the RECAP Infrastructure Optimiser adds value by augmenting the application optimisation with additional infrastructure-specific information that can encompass the business objectives of both application and infrastructure providers.

In this chapter, we detailed the infrastructure optimisation architecture tasked with establishing this more holistic optimisation recommendation. Section 4.3 outlined the problem formulation describing the varied and detailed inputs required for optimisation that must be mathematically and programmatically described and traversed, including infrastructural components, the application, and constraints to optimisation. This was followed by a description of the models that inform optimisation, and the evolutionary algorithm and utility functions used to mathematically and programmatically select from sub-optimal solutions.

The value in the described approach is difficult to articulate and visualise. When one considers the size and complexity of modern hyperscale architecture, it is apparent that such a granular, mathematical, and programmatically implementable approach is required in order to extract value from the nuanced and humanly incomprehensible myriad of available options.

References

Jin, Hao, Deng Pan, Jing Xu, and N. Pissinou. 2012. *Efficient VM Placement with Multiple Deterministic and Stochastic Resources in Data Centers.* Proceedings of the IEEE Global Communications Conference (GLOBECOM), 2505–2510.

Li, X., A. Ventresque, J. Murphy, and J. Thorburn. 2014. *A Fair Comparison of VM Placement Heuristics and a More Effective Solution.* Proceedings of the 13th IEEE International Symposium on Parallel and Distributed Computing (ISPDC), 35–42.

Loomba, R., T. Metsch, L. Feehan, and J. Butler. 2017. *Utility-Driven Deployment Decision Making.* Proceedings of the 10th International Conference on Utility and Cloud Computing, Austin, TX, 207–208.

Loomba, Radhika, Keith A. Ellis, Paolo Casari, Rafael García, Thang Le Duc, Per-Olov Östberg, and Johan Forsman. 2019. *Final Infrastructure Optimisation and Orchestration*. RECAP Deliverable 8.4, Dublin, Ireland.

Meng, X., V. Pappas, and L. Zhang. 2010. *Improving the Scalability of Data Center Networks with Traffic-aware Virtual Machine Placement*. IEEE INFOCOM, San Diego.

Metsch, T., O. Ibidunmoye, V. Bayon-Molino, J. Butler, F Hernández-Rodriguez, and E. Elmroth. 2015. *Apex Lake: A Framework for Enabling Smart Orchestration*. Proceedings of the Industrial Track of the 16th International Middleware Conference, Vancouver.

RECAP Consortium. 2018. Deliverable 6.1. Initial Workload, Load Propagation, and Application Models.

Superfluidity Consortium. 2017. Deliverable 5.1. Function Allocation Algorithms Implementation and Evaluation.

Xu, J., and J.A.B. Fortes. 2010. *Multi-objective Virtual Machine Placement in Virtualized Data Center Environments*. Proceedings of IEEE/ACM International Conference on Green Computing and Communications (GreenCom) & International Conference on Cyber, Physical and Social Computing (CPSCom), Hangzhou, 179–188.

CHAPTER 5

Simulating Across the Cloud-to-Edge Continuum

*Minas Spanopoulos-Karalexidis, Christos K.
Filelis Papadopoulos, Konstantinos M. Giannoutakis,
George A. Gravvanis, Dimitrios Tzovaras,
Malika Bendechache, Sergej Svorobej,
Patricia Takako Endo, and Theo Lynn*

Abstract As growth and adoption of the Internet of Things continue to accelerate, cloud infrastructure and communication service providers (CSPs) need to assure the efficient performance of their services while meeting the Quality of Service (QoS) requirements of their customers and

M. Spanopoulos-Karalexidis • C. K. Filelis Papadopoulos • K. M. Giannoutakis
• D. Tzovaras
Information Technologies Institute, Centre for Research and Technology Hellas,
Thermi, Greece
e-mail: mspanopoulos@iti.gr; kgiannou@iti.gr; dimitrios.tzovaras@iti.gr

G. A. Gravvanis
Democritus University of Thrace, Xanthi, Greece
e-mail: ggravvan@ee.duth.gr

their end users, while maintaining or ideally reducing costs. To do this, testing and service quality assurance are essential. Notwithstanding this, the size and complexity of modern infrastructures make real-time testing and experimentation difficult, time-consuming, and costly. The RECAP Simulation Framework offers cloud and communication service providers an alternative solution while retaining accuracy and verisimilitude. It comprises two simulation approaches, Discrete Event Simulation (DES) and Discrete Time Simulation (DTS). It provides information about optimal virtual cache placements, resource handling and remediation of the system, optimal request servicing, and finally, optimal distribution of requests and resource adjustment, with the goal to increase performance and concurrently decrease power consumption of the system.

Keywords Simulation • Discrete event simulation • Discrete time simulation • Resource allocation • Capacity planning • CloudSim • CloudLightning Simulator • Distributed clouds

5.1   INTRODUCTION

In this chapter, an overview of the RECAP Simulation Framework which comprises two different simulation approaches is outlined and discussed— Discrete Event Simulation (DES) and Discrete Time Simulation (DTS). The RECAP Simulation Framework offers two approaches to recognise the characteristics, requirements, and constraints of cloud and

M. Bendechache • S. Svorobej (✉)
Irish Institute of Digital Business, Dublin City University, Dublin, Ireland
e-mail: Malika.Bendechache@dcu.ie; sergej.svorobej@dcu.ie

P. T. Endo
Irish Institute of Digital Business, Dublin City University, Dublin, Ireland

Universidade de Pernambuco, Recife, Brazil
e-mail: patricia.endo@upe.br

T. Lynn
Irish Institute of Digital Business, DCU Business School, Dublin, Ireland
e-mail: theo.lynn@dcu.ie

communication service providers. As such, the RECAP Simulation Framework offers a solution for (1) SMEs and large hyperscale cloud and network operators, and (2) providers requiring rapid less-detailed simulation results and those requiring a more-detailed simulation.

DES focuses on aggregating each incoming request in the form of events, regarding their entry timestamp, and usually in a pipelined manner. These events are stored in an initialised list of tasks, retained in memory and augmented with each incoming task. In order to accommodate all this information, the required resources and especially memory requirements are significant and large. Thus, DES is suitable for simulating smaller and intensively detailed scenarios, in order to maintain accuracy at high levels. DTS on the other hand provides the potential to simulate larger scenarios with its ability to scale up significantly. This is feasible due to the fact that DTS does not need precomputation and storage of future events; it uses a time-advancing loop, where the requests are entering the system in respective time steps during the simulation. This results in a significant reduction in memory requirements, providing significant improvements in the ability to scale up the simulation. DTS does not offer the level of detail of DES, but it can be a useful and accurate tool for simulating real large-scale scenarios, while maintaining resource consumption on reasonable levels.

In this chapter, a high-level overview of the RECAP Simulation Framework is presented and discussed. This is followed by a brief overview of the RECAP DES framework, followed by a short case study illustrating its applicability for cloud infrastructure and network management. Then, the RECAP DTS framework is presented with a short case study illustrating its applicability for simulating virtual content distribution networks.

5.2   HIGH-LEVEL CONCEPTUAL OVERVIEW OF THE RECAP SIMULATION FRAMEWORK

The RECAP Simulation Framework facilitates reproducible and controllable experiments to support the identification of targets for the deployment of software components and optimising deployment choices before actual deployment in a real cloud environment. It was designed specifically to simulate distributed cloud application behaviours and to emulate data centre and network systems across the cloud-to-thing continuum.

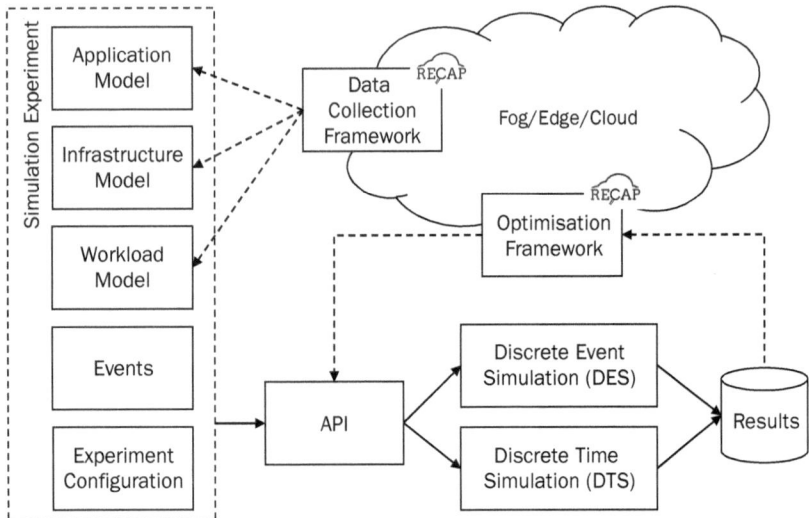

Fig. 5.1   High-level conceptual overview of the RECAP simulation framework

Figure 5.1 presents a high-level conceptual overview of the RECAP simulation framework comprising the following components: Simulation Experiment Models, Application Programming Interface (API), Simulation Manager, and Model Mappers for DES and DTS simulators.

The system monitoring data obtained from the RECAP Data Analytics Framework (presented in Chap. 2) are used to compile the Simulation Experiment Models:

- Application Model: represents the application components and their connections and behaviour, i.e. application load propagation and operational model;
- Infrastructure Model: describes the physical infrastructure (network topology and (physical and virtual) machines' configurations) where the application will be hosted;
- Workload Model: describes how the workload generated by the users is distributed and processed by the application components; and,
- Experiment Configuration: where the simulation user configures the simulation parameters, such as simulation time, simulation time step, log files, and input files.

The RECAP Optimisation Framework makes use of the RECAP Simulation Framework to evaluate different application deployments and infrastructure management alternatives in terms of cost, energy, resource allocation, and utilisation, before actuating on real application deployments. This integration is done through an API that receives the models, the experiment configuration, and the set of simulation scenarios, and sends them to a web-based REST API. Depending on the type of API call, the experiment is forwarded to the RECAP DES or DTS simulator. Once simulation is completed the results can be accessed from the chosen storage method, e.g. local CSV files or a database.

5.3   DISCRETE EVENT SIMULATION

5.3.1   Overview

Discrete Event Simulation (DES) is a system modelling concept wherein the operation of a system is modelled as a chronological sequence of events (Law et al. 2000). DES-based decision support processes can be divided into three main phases: modelling, simulation, and finally, results analysis. During the modelling phase, a simulated system is defined by grouping interacting entities that serve a particular purpose together into a model. Once the representative system models are created, the simulation engine orchestrates a time-based event queue, where each event is admitted to the defined system model in sequence. An event represents actions happening in the system during operation time. Depending on the event type, the system reaction is simulated, and associated metrics captured. These metrics are collected at the end of the simulation for results analysis. Therefore, system behaviour can be examined under different conditions. Using DES is beneficial in a complex real non-deterministic small-to-medium-sized system environment (SME) where the system definition using mathematical equations may no longer be a feasible option (Idziorek 2010).

5.3.2   The RECAP DES Framework

The RECAP DES Framework captures system configurations by using Version 3 of the Google Protocol Buffers technology.[1] This implementation approach was chosen to ensure model schema would remain

---

[1] https://developers.google.com/protocol-buffers/

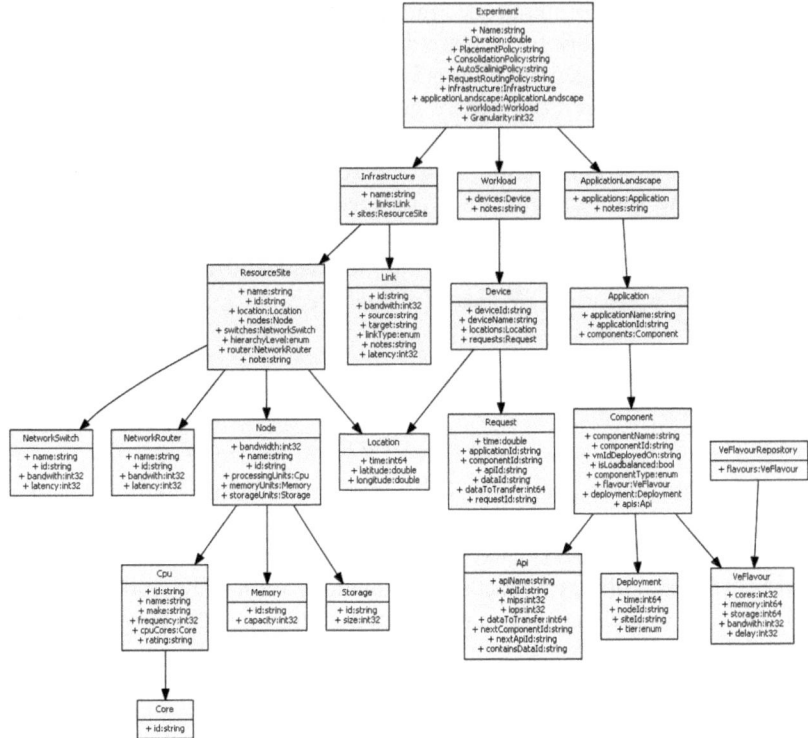

Fig. 5.2 DES simulation model data format (inputs)

programming language-neutral allowing serialised models to be used across multiple language platforms such as Java, C++, C#, or Python. In addition, Protocol Buffers are simple and faster to use and are smaller in size compared to XML or JSON notations. Speed and file size of the model are important when dealing with large-scale systems by managing the memory footprint of simulation framework and sending data over a RESTful web API client.

As shown in Fig. 5.2, the RECAP DES simulator root element is the *Experiment* class which contains nested system models, the name identifier of the simulation experiment, and its parameters used by the simulation engine, i.e. *Duration, Granularity, PlacementPolicy, ConsolidationPolicy,*

AutoScalingPolicy and *RequestRoutingPolicy*. The *Duration* parameter defines the length of the simulation experiment in simulated time while the *Granularity* is a multiplier for the number of requests represented by a single simulation event. Placement, consolidation autoscaling, and request routing policies are optional attributes which specify the name of any resource management policies which can be integrated within the simulator. In addition, the *Experiment* class contains nested *Infrastructure*, *Workload*, and *ApplicationLandscape* models, which describe edge to cloud system composition and behaviour.

The *Infrastructure* model captures the hardware characteristics of a distributed network and computes hardware locations. Each *ResourceSite* component in the model represents a virtualised cloud/edge/fog data-centre location which is geographically distributed with *Location* class containing latitude and longitude spatial information. Nested *NetworkSwitch*, *NetworkRouter* and *Node* model components capture network bandwidth, latency, and compute resource (CPU, memory, storage) capacity at each location.

The *Workload* model contains mappings between devices and requests devices made to the system. The *Device* system component has *Name* and *ID* attributes as well as a time-dependent *Location* array and an array of *Requests*. Each *Request* component describes a request of the device (user) made to the system at a specific geographical location. The *Request* attributes capture the time of the request, amount of data to transfer, type of data, and application model API where the request is destined for.

The *ApplicationLandscape* component contains information on the applications running in the virtualised infrastructure. Each application can be composed of multiple interconnected components, and each application component can have multiple functions expressed through an API definition; hence in the model, we have *Application*, *Component*, and *API* classes describing the relationships. The model assumes a one-to-one relationship between the application component and a Virtual Entity (VM or Container) it is deployed to. Therefore, the *Component* class also contains the *Deployment* class describing which hardware node it is deployed to and a *VeFlavour* class specifying what resources it requires.

The simulation results, called outputs, are also arranged in a structured form using Protocol Buffers. The proposed format structure is captured within a class diagram shown in Fig. 5.3.

The *ExperimentResult* root class splits into two arrays of simulated system behaviour metrics: *ResourceSiteMetrics* and *ApplicationMetrics*. As the

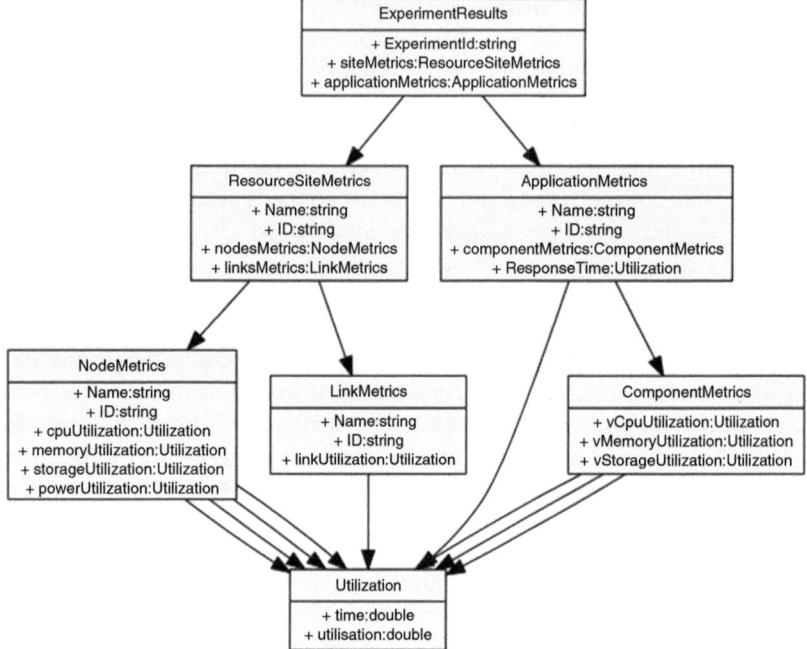

Fig. 5.3   DES simulation results format (outputs)

name suggests the *ResourceSiteMetrics* class contains information on hardware utilisation and *ApplicationMetrics* contains information on application performance metrics. The subclass *NodeMetrics* stores per-node metrics of CPU, memory, storage, and power utilisation. The subclass *LinkMetrics* stores each link utilisation bandwidth. Similarly, the subclass *ComponentMetrics* captures utilisation of resources per individual virtual entity besides including a response time metrics for end-to-end application performance in the upper *ApplicationMetrics* class. All of the measurements are captured at regular time intervals, hence attributes in the *Utilisation* class list time and the actual utilisation value. The *Utilisation* class is built as an abstract and can be extended to fit different types of measurements.

Once the models are created based on the desired system parameters, they can be then loaded into the RECAP DES Simulator. The RECAP

DES Simulator is based on CloudSim[2] with a custom DES implementation in the back-end. To load, read, and query the simulation input models and output results shown in Figs. 5.2 and 5.3, Google Protocol Buffers library provides auxiliary methods ensuring ease of use.[3]

5.3.3   Cloud Infrastructure and Network Management: A RECAP DES Framework Case Study

To illustrate how the RECAP simulation and modelling approach can be used by communication service providers, we present the application of the RECAP Simulation Framework for mobile technology service management within fog/cloud computing infrastructure. This case study is based on automated services and infrastructure deployment (using virtual network functions (VNF)), automated orchestration, and optimisation services to reach the desired QoS for the different network services. We model distributed infrastructure and a VNF service application chain using the RECAP DES Framework.

5.3.3.1   Infrastructure Model

The infrastructure simulation model was designed and implemented to capture available physical characteristics of real edge infrastructure and used input from the infrastructure models described in Chap. 4. It consists of several sites that are interconnected by links between each other. Each *Site* entity in the model represents a location that is hosting network and/or computing equipment, such as switches, routers, and computing nodes. *NetworkSwitch* and *NetworkRouter* capture attributes of bandwidth and latency while *Nodes* in addition to bandwidth also capture properties of CPU, Memory, and Storage. For the simulation experiments, physical infrastructure for 45 distributed sites was modelled; each site contains a router for handling inbound and outbound internet traffic and two switches handling control plane and user plane traffic separately. This meant that any traffic that is received or transmitted from the site is traversing through the router and internal traffic between physical hosts and is flowing through routers only. The user plane switch was assigned 40 Gbps bandwidth and control plane switch 1 Gbps where routers were

---

[2] http://www.cloudbus.org/cloudsim/
[3] The methods are well documented in tutorials widely available for a range of programming languages: https://developers.google.com/protocol-buffers/docs/tutorials

assigned 100 Gbps. Bandwidth assumptions were made based on the data gathered from testbed experiments and correspond to the volume of traffic observed. In addition, links between sites were assigned additional latency delays proportionate to the distance between locations; hence, requests sent between sites take more time to arrive.

5.3.3.2 *Application and Workload Propagation Model*

Application behaviour for the simulation is realised by implementing a modelling concept that captures data flow through multiple interconnected, distributed, components. Each component is represented as a virtual entity (VM or container) that is assigned to a physical machine in a site and has access to the portion of resources.

Application behaviour logic is realised through multiple interconnected API elements that each component has. The API represents a model object that holds information on resource demand and connection to the next component in line, thus forming a logical path between different application components. The current use case is based around NFV paradigm, and the used VNF chain is of a virtualised LTE stack which consists of user data plane and control data plane virtual components:

- eNodeB user plane denoted as *CU-C*
- Mobility Management Entity control plane denoted as *MME-C*
- eNodeB control plane denoted as *CU-U*
- Serving Gateway-User plane denoted as *SGW-U*
- Packet Data Network Gateway-User plane denoted as *PGW-U*

For example, Fig. 5.4 graphically describes one of the possible application topologies where application components like CU-C, CU-U, SGW-U, and PGW-U are located on one site and component MME-C is located on another site. In this example, both user plane download (green) and upload (red) requests are executed on one site, but the control plane (blue) requests require to travel to another site to be processed resulting in longer processing delays. When a request arrives at a component, based on the API parameters number of resources are requested from the hardware to process this request. Once the appropriate amount of resources is available, the request is sent further in the system according to the API connection path. During the simulation experiments, 1482 different application configurations (placements) were generated and combined with the infrastructure model and corresponding workload models.

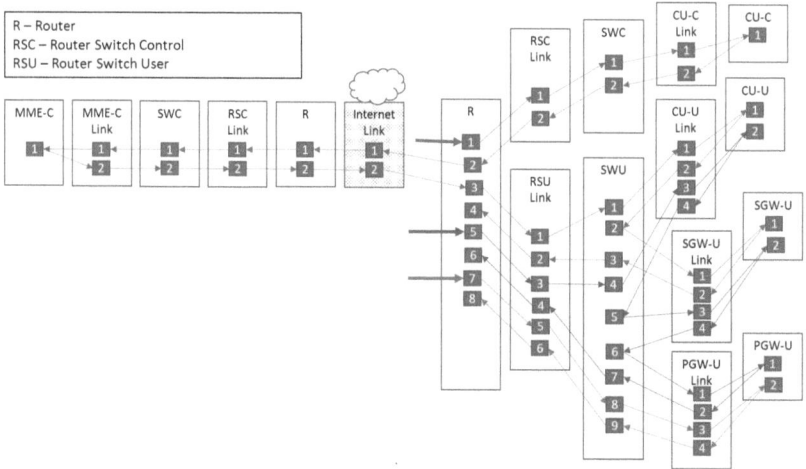

Fig. 5.4 Application simulation model example

The workload simulation model was implemented to capture the number of end-user devices that use the system for the duration of the simulation experiment. The model contains an array of devices each containing multiple requests. Furthermore, each request contains information on the arrival time, the size of the request, and application information for its destination. The number of users varies depending on the VNF placement on the infrastructure sites. More densely populated areas have more users, and this aspect is reflected in the workload models; hence, each placement experiment has a bespoke number of users. User-request parameters were based on the data gathered from testbed experiments, and an average of the quantity and size of the request was done, along with defining them into three categories *User Download, User Upload*, and *Control*. As shown in Table 5.1, on average data download request in user plane is 13,927 bits and user makes 2808 requests per hour. For data upload requests user sends around 224 requests per hour each of 8572 bits in size and finally control plane administrative requests were taken as a fracture of upload traffic and amount to 6 requests per hour each 219 bits in size.

DES simulations were executed in concurrent batches of 5 parallel runs on a dedicated VM in a testbed. The VM configuration was set to 8 CPU cores, 64 GB of RAM, and 500 GB attached volume storage. Each

Table 5.1 Average values of user requests

Request type	Requests per user per hour	Request size (bits)
User download	2808	13,927
User upload	224	8572
Control	6	219

simulation experiment was set for **3600**-second duration of simulated time and on average took around **30** seconds of wall clock time to complete.

The simulation results were analysed using utility functions with the total resource utilisation and the cost of the allocated machine serving as the provider utility with equal weighting, along with the network bandwidth consumed and total latency serving as the customer utility with equal weighting. The mathematical formulation of these utility functions is available in Chap. 4. Our goal was to minimise resource utilisation utility, latency utility, and cost utility while maximising network bandwidth utility. The total utility for the placement was then defined as an equally weighted sum of normalised provider and customer utility.

5.3.3.3 RECAP DES Results

Results showed that by fixing the provider utility or customer utility, there is further scope to maximise the corresponding utility by changing the placement distribution of the VNFs. This is highly beneficial for stakeholders when making business decisions regarding the available infrastructure. These decisions map back to the values considered within the definition of both utility functions such as response time, bandwidth, latency, and utilisation costs.

5.4   Discrete Time Simulation

5.4.1   Overview

Discrete Time Simulation (DTS) is a simulation technique based on a time-advancing loop of predefined starting and ending time. The defined time step is a portion of time values (usually seconds) that the user has the ability to set before the execution. During each time step, potential new requests/events enter the system from the defined entry points. The major advantage of DTS is that no precomputation and storage of future events are needed, thus resulting in a significant reduction in memory consumption requirements. This also suggests the possibility to dynamically allocate simulated resources based on current computational load.

5.4.2   The RECAP DTS Framework

The RECAP DTS simulation framework is based on the CloudLightning Simulation Platform, which is designed to simulate hyperscale heterogeneous cloud infrastructures (Filelis-Papadopoulos et al. 2018a, b). It was built using the C++ programming language utilising OpenMP to exploit parallelism and acceleration in computations where applicable. The RECAP Simulation Framework focuses on optimally placing VMs as caches or containers in a network while taking into account efficient resource utilisation, reduction of energy consumption and end-user latency, and load balancing for minimisation of network congestion.

The CloudLightning Simulation Platform was developed to simulate hyperscale environments and efficiently manage heterogeneous resources based on Self-Organisation Self-Management (SOSM) dynamic resource allocation policies. The simulated cloud architecture is based on the Warehouse Scale Computer (WSC) architecture (Barroso et al. 2013). It manages to maintain a simplistic approach by utilising models that do not demand extremely high computational effort and, at the same time, maintain accuracy at adequate levels. The utilisation of a time advancing loop, rather than a discrete sequence of events, enables the potential to use these dynamic resource allocation techniques while also providing high scalability due to the lack of restrictions in memory requirements.

A brief summary of the basic characteristics of the CloudLightning Simulation architecture is as follows. The gateway lies at the topmost level on the master node and the cells, which are connected directly to the gateway, hosted on separate distributed computing nodes at a lower level. Each cell is responsible for the underlying components, such as cell's broker, network, telemetry, and finally, hardware resources. Key responsibilities of the gateway are (1) communication with the available cells, in the essence of data transport, fragmentation, and communication of the task queues between the cells with the appropriate load balancing on each time step; and (2) receiving and maintaining metrics and cells' status, amongst others. From the cell's perspective, the key responsibilities are (1) the aforementioned communication with the gateway, including simulation parameters and initialisation of the underlying components, and additionally, sending status and metrics' information to the gateway; and (2) task queue receipt on each time step, finding the optimal component with the required available resources, utilising the SOSM engine, and finally, executing the tasks.

Considering the above, many CloudLightning Simulator components were adopted for the RECAP Simulation Framework including power consumption modelling, resource utilisation (vCPU, memory, storage), and bandwidth utilisation. Some of the most important differences between the two frameworks are: (1) the focus of the RECAP Simulator which is on the optimal cache placement in the network, and (2) the difference in task servicing, and specifically in task deployment to the available nodes. The CloudLightning Simulator utilises a Suitability Index formula and is based on the required weights communicated by the gateway to the underlying components. The most appropriate node is assigned with the incoming task, by adopting a first-fit approach. The RECAP Simulator, on the other hand, utilises caches with the corresponding content placed in the network. In order to assign a task to a respective node, it performs a search for the optimal available node, which offers, in excess, the required resources while adequately handling network congestion. Experimental results of the CloudLightning Simulator demonstrated that it can accurately handle simulations of hyperscale scenarios with relatively low computational resources. This is particularly suitable for large distributed networks that many Tier 1 network operators manage.

5.4.3   Network Function Virtualisation—Virtual Content Distribution Networks: A RECAP DTS Case Study

Traditional Content Distribution Network (CDN) providers occasionally install their hardware, such as customised hardware caches, in third-party facilities or within the network of an Internet Service Provider. BT has such a scenario, which the RECAP DTS simulator utilises as a case study. BT's main activities focus on the provision of fixed-line services, broadband, mobile and TV products and services, and networked IT services as well. BT hosts customised hardware caches from the biggest CDN operators in their network. Considering the fact that it would be extremely hard, for sensitive reasons, for content providers to install their hardware in many locations across the UK in the edge nodes of BT's network (also known as Tier 1 MSANs (Multi-Service Access Nodes)), there lies the need to provide an alternate solution. In order to ensure the required QoS for their virtual network functionalities, the introduction of a Virtual CDN (vCDN) provides a beneficial approach, which aims to replace the presence of multiple physical caches in the network, with standard servers and storage providing multiple virtual applications per CDN operator. BT accomplishes that, by installing the appropriate compute infrastructure at its edge nodes (MSANs) and thus offering a CDN-as-a-Service (CDNaaS). In this way, operational costs are significantly declined and additionally, the content is stored in virtual caches closer to end user, thus minimising end-user latency and maximising user experience.

5.4.3.1 *DTS Architecture and Component Modelling*

The topological architecture is divided in four tiers, in a hierarchical order namely MSAN, Metro, Outer-Core, and Inner-Core, in a total number of 1132 nodes; more information on infrastructure architecture is provided in the next subsection. Note that in order to maintain efficient simulation accuracy, a specific time interval is selected, at which all the components update their status. This provided the opportunity to reduce computational cost, but it is essential to mention that the choice of the interval value is critical. A small interval can lead to huge computational effort and reduce performance, while a large interval can lead to major accuracy leaks, considering that whole requests could be missed during the status update process. Considering all these, the RECAP DTS framework provides the essential scalability for the current use case. The ambition is to improve the efficiency of vCDNs systems by replacing multiple customised physical caches running multiple virtual applications per CDN operator.

Figure 5.5 depicts the DTS architecture optimised to simulate a vCDN network. The **Graph Component** is responsible for the input topology of the simulation, which is fed to the component as an input file in Matrix Market storage format. The structure is stored as a Directed Acyclic Graph (DAG) in the component, in the form of a sparse matrix, with the number of rows being the total number of sites and each row, the ID of a site. More specifically, it is stored both as a Compressed Sparse Row (CSR) and Compressed Sparse Column (CSC) format which results in a faster traverse of the available connections. These connections are indicated by off-diagonal values and point to links with lower level sites, while the diagonal values denote the level/tier of the respective site.

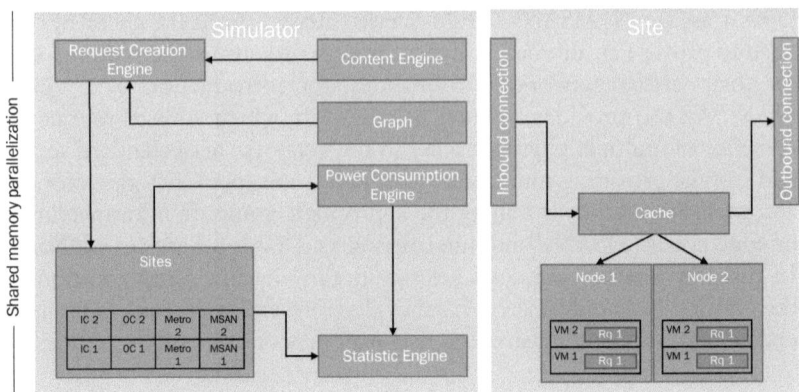

Fig. 5.5   DTS architecture

All the available sites are retained in a *vector of sites*, where a site is an object of the Site class which contains a number of attributes. Each site has a unique *ID* value and a *type* value, indicating the tier of the respective site. Furthermore, a two-dimensional vector retains the available connections *input* and *output* to the immediate upper and lower level respectively. Note that in case of the first level there are no input connections, and similarly, there are no output connections in case of the last level. Another vector retains the *output bandwidth* of the output connections as double precision values (Gbps). All the attached nodes, of predefined type, to the site are also retained in a vector and contain resource information such as *CPU*, *Memory*, and *Storage*. These nodes are mostly utilised by the power consumption component. In addition to nodes, there is a list of the hosted VMs deployed to the nodes of each site and provide same information with the addition of *Network*. Additionally, a map, which contains the *cached content* in the VMs, is used in order to simplify and speed up the search of specific content type or available VMs. All the *cache hits and misses* are also stored into a vector; these refer to all active VMs hosted by a single site. Lastly, a site can *forward requests* to sites at a higher tier, due to lack of VMs or insufficient resources in them. These forwarded requests are also retained in a list and have an impact only on the network bandwidth of the site.

The *Content Component* retains all the potential information referring to the type of content a VM in a site can serve. Specifically, this information includes minimum and maximum duration of each type of content and the requirements that need to be met by the VM in order to serve it. Also, the probability of a cache hit for each type of content and maximum number of requests that can be served are included. The requirements of each type of cached content request is provided by the ratio of the VM require- ments of this specific request to the aforementioned maximum number of requests the VM can serve.

The requests are generated by the **Request Creation Engine**, which is responsible for the insertion of a group of requests to the system in each time step. This component is based on a uniform distribution generator, which produces requests of each type of content and duration between a given interval denoting the minimum and maximum requests permitted in a time step. Each request contains the following information: *duration* of the request, *type* of content, and the *site* from which it enters the system. For each of the inserted requests, a path (list of sites) is formed showing the flow the content will follow in order to reach the user. During path creation, each site traversed and is appended to the end of the list; this continues until a cache hit occurs or otherwise. If the last element is not a cache hit, the request is rejected. When a cache hit occurs, the content flows downwards from that site to all sites of the path of a lower tier until it reaches the user.

During each time step, the duration of all requests is reduced until it reaches zero, at the point they are considered served and can be discarded from the system, thus freeing up resources. This procedure takes place in each site as well as in the site's nodes and VMs that update their status. This is where OpenMP provides acceleration of the computations on shared memory systems. Each site is independent; thus, they can be assigned to the available threads and their status updated without any interference. During each time step, each site checks its current status, duration of requests, any new additions, or any finished, and respectively adjusts available resources. This procedure has no data traces, as sites are independent. Thus, this computational-intensive process, considering the number of sites and the huge number of time steps can be performed in parallel, saves significant amount of time and increases performance. Apart from status update, at each time step, another component is also used, the **Power Consumption Component**. This calculates the power/energy con- sumption of the site's nodes depending on their type (Makaratzis et al. 2018).

Power consumption along with other metrics is stored in the last component, the ***Statistics Engine*** which is deployed at specific time intervals. It contains all metrics of the vCDN network and each site generally. Metrics include the aforementioned power consumption, cache hits and misses, and other stats per level such as cumulative accepted and rejected requests per level, average vCPU, Memory, Storage, and Network utilisation per level. The Statistics Engine outputs these metrics to files at each specific time interval. Note again that accurate selection of this interval is critical otherwise it can lead to either huge writing effort (in the case of a small interval) or under-sampling (in the case of a large interval).

5.4.3.2 Infrastructure Model

The considered vCDN system is hierarchical and has sites located at four different levels: (1) inner core, (2) outer-core, (3) metro, and (4) Multi-Service Access Node (MSAN), as illustrated in Fig. 5.6.

The physical network topology is composed of 1132 sites. Each site can host physical machines (nodes) that, in turn, can host vCDNs containing content requested by customers. Moreover, each site has predefined upload and download bandwidth as well as inbound and outbound connections. The general structure of a site is given in Fig. 5.7.

Each node inside a site can host multiple VMs and each VM services specific content. However, multiple VMs can service the same content if it

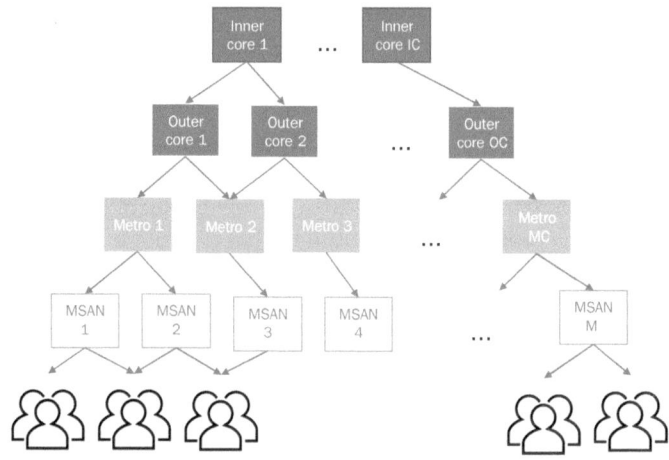

Fig. 5.6 BT hierarchical level of sites

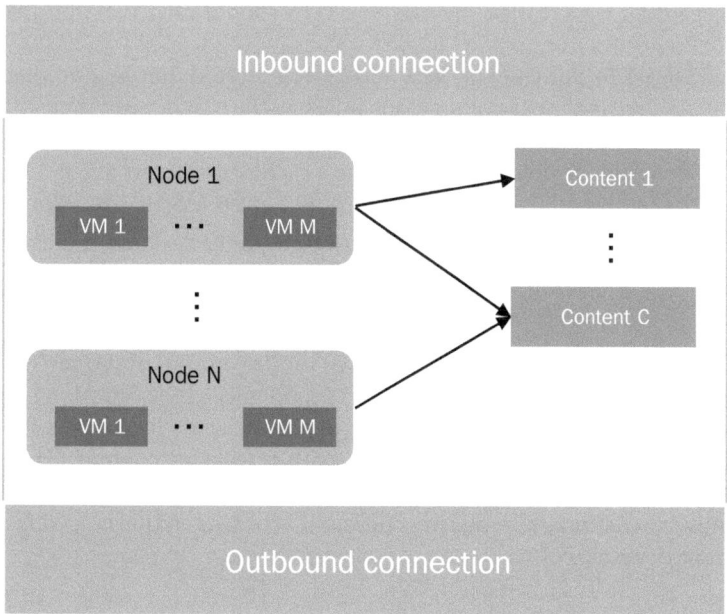

Fig. 5.7 A site architecture of DTS

is requested by a large number of users, since each VM has predefined capacity to service customer.

5.4.3.3 *Application and Workload Propagation Model*
Each type of content (content with different sizes) characterises a VM. For example, content of type j is serviced by a specific VM with predefined characteristics described above. Thus, each type of content has its unique accompanying requirements. Moreover, each user of a specific content requires predefined bandwidth and occupies the system for a variable amount of time lying between predefined intervals. For each type of content, the requirements in terms of VMs as well as per user bandwidth are defined. The same content can be hosted in several VMs on the same site, since the number of users requiring a specific type of content serviced by a VM is limited. The characteristics of content include required vCPUs per VM, required memory per VM, required storage per VM, maximum number of customers per VM, network bandwidth required per VM at full capacity.

Regarding the creation of VMs, the DTS should offer two options:

1. **Static**: In this case, no new VM can be created during a simulation. This requirement comes from Infrastructure Optimiser that will use the RECAP Simulation Framework to decide placements; or
2. **Dynamic**: In this case, every time a cache miss occurs at MSAN site, the requested content is copied to that given MSAN site. Moreover, when there is no user requesting a given content, this content (VM) is deleted from the node. If the node has no more available resources, then the request is rejected.

The Request Creation Engine (RCE) creates a series of requests based on random number generators following a preselected distribution such as Uniform, Normal, or Weibull. Each request performed by a user is considered to have predefined requirements with respect to content. Thus, all customers of a certain content type require the same amount of resources. However, customers requiring different content require a different amount of resources.

5.4.3.4 RECAP DTS Results

The results of this simulation are presented in detail in C. K. Filelis-Papadopoulos et al. (2019). In summary, we find that parallel performance (status update) is significantly increasing proportionally to the number of requests. In addition, resource consumption seems to reach stability, for all levels, by the time initial requests have finished execution. The lowest level contributes the most to resource and node underutilisation through request forwarding to upper layers as a result of probabilistic caching and single VM hosting. This leads to reduced active server utilisation and increased power consumption concurrently due to node underutilisation. Nevertheless, energy consumption is improved with the reduction of VMs on the lowest level; thus, these sites act as forwarders to the immediate upper layer. This impacts efficient resource utilisation in the upper layers, while they service more requests forwarded from the bottom tier.

Other experiments focused on different probabilities for a cache hit, such as 0.4 and 0.8. The former leads to requests servicing from the topmost layer due to the fact that content requested is not potentially cached in the lower layers and thus a cache miss occurs and the requests are forwarded. The latter increases the probability for a cache hit to occur and thus requests are serviced mostly from lower layers. More specifically, the intermediate level serves a significantly increased amount of tasks. Note that, with 0.8 probability the energy consumption is considerably decreased when compared to the other two cases. Nevertheless, high probability denotes that the content is cached in a great portion of the distributed caches in the network, as the probability value acts as a mechanism to transfer workload between the corresponding nodes and tiers of the network. Thus, potential deployment of virtual caches of specific content in great numbers could result in higher costs and storage requirements. On the other hand, lower probability denotes a significant reduction in virtual cache numbers, especially on lower levels. As discussed earlier, this results in higher service rates from the top layers and furthermore in potential network congestion due to increased data traffic in the links of the network, request rejection, and increased end-user latency in request servicing from nodes significantly further from the end users.

Finally, we performed experiments with an increased number of levels. The scalability performance results suggest that the simulator scales linearly with the number of input requests, considering the major increase (mostly two times) in memory requirements. The results illustrate that the framework is capable of executing large-scale simulations in a feasible time period even with significant memory requirements (as number of threads increases, the need of memory for local data storage also increases) and at the same time maintaining required high levels of accuracy. Thus, the RECAP DTS framework can be a useful tool for content providers to validate their overall performance.

5.5   CONCLUSION

In this chapter, the RECAP Simulator Framework, comprising two simulation approaches—DES and DTS, was presented. The design and implementation details of the RECAP simulation framework were given in both simulation approaches, coupled with case studies to illustrate their applicability in two different cloud and communication service provider use cases. The main advantage of this framework is the fact that depending on the target use case requirements, an appropriate simulation approach can be selected based on a time-advancing loop or a discrete sequence of events. Thus, by providing this flexibility, focus can be given on the level of accuracy of the results (DES) or the scalability and dynamicity (DTS) of the simulation platform.

From the experimentation performed, the RECAP simulation platform was capable of efficiently simulating both discrete event and discrete time use cases thus providing a useful tool for non-data scientists to forecast the placement of servers and resources by executing configurable prediction.

REFERENCES

Barroso, L.A., J. Clidaras, and U. Hoelzle. 2013. *The Datacenter as a Computer: An Introduction to the Design of Warehouse-Scale Machines.* Morgan & Claypool. https://doi.org/10.2200/S00516ED2V01Y201306CAC024.

Filelis-Papadopoulos, C.K., K.M. Giannoutakis, G.A. Gravvanis, and D. Tzovaras. 2018a. Large-scale Simulation of a Self-organizing Self-management Cloud Computing Framework. *The Journal of Supercomputing* 74 (2): 530–550. https://doi.org/10.1007/s11227-017-2143-2.

Filelis-Papadopoulos, C.K., K.M. Giannoutakis, G.A. Gravvanis, C.S. Kouzinopoulos, A.T. Makaratzis, and D. Tzovaras. 2018b. Simulating Heterogeneous Clouds at Scale. In *Heterogeneity, High Performance Computing, Self-organization and the Cloud,* 119–150. Cham: Palgrave Macmillan.

Filelis-Papadopoulos, Christos K., Konstantinos M. Giannoutakis, George A. Gravvanis, Patricia Takako Endo, Dimitrios Tzovaras, Sergej Svorobej, and Theo Lynn. 2019. Simulating Large vCDN Networks: A Parallel Approach. *Simulation Modelling Practice and Theory* 92: 100–114. https://doi.org/10.1016/j.simpat.2019.01.001.

Idziorek, Joseph. 2010. *Discrete Event Simulation Model for Analysis of Horizontal Scaling in the Cloud Computing Model.* Proceedings of the 2010 Winter Simulation Conference, 3003–3014. IEEE.

Law, Averill M., W. David Kelton, and W. David Kelton. 2000. *Simulation Modeling and Analysis.* New York: McGraw-Hill.

Makaratzis, Antonios T., Konstantinos M. Giannoutakis, and Dimitrios Tzovaras. 2018. Energy Modeling in Cloud Simulation Frameworks. *Future Generation Computer Systems* 79 (2): 715–725. https://doi.org/10.1016/j.future.2017.06.016.

CHAPTER 6

Case Studies in Application Placement and Infrastructure Optimisation

Miguel Angel López-Peña, Hector Humanes, Johan Forsman, Thang Le Duc, Peter Willis, and Manuel Noya

Abstract This chapter presents four case studies each illustrating an implementation of one or more RECAP subsystems. The first case study illustrates how RECAP can be used for infrastructure optimisation for a 5G network use case. The second case study explores application

M. A. López-Peña (✉) • H. Humanes
Sistemas Avanzados de Tecnología, S.A. (SATEC), Madrid, Spain
e-mail: miguelangel.lopez@satec.es; hector.humanes@satec.es

J. Forsman • T. Le Duc
Tieto Product Development Services, Umeå, Sweden
e-mail: johan.forsman@tieto.com; thang.leduc@tieto.com

P. Willis
BT plc, London, UK

M. Noya
Linknovate, Palo Alto, CA, USA
e-mail: manuel@linknovate.com

© The Author(s) 2020   117
T. Lynn et al. (eds.), *Managing Distributed Cloud Applications and Infrastructure*, Palgrave Studies in Digital Business & Enabling Technologies, https://doi.org/10.1007/978-3-030-39863-7_6

optimisation for virtual content distribution networks (vCDN) on a large Tier 1 network operator. The third case study looks at how RECAP components can be embedded in an IoT platform to reduce costs and increase quality of service. The final case study presents how data analytics and simulation components, within RECAP, can be used by a small-to-medium-sized enterprise (SME) for cloud capacity planning.

Keywords Infrastructure management • Network management • Network functions virtualisation • Virtual content distribution networks • Smart cities • Capacity planning • Application optimisation • Infrastructure optimisation • Big Data analytics • 5G networks

6.1   Introduction

This chapter illustrates how RECAP's approach to the management and optimisation of heterogeneous infrastructure across the cloud-to-edge spectrum can address challenges to a wide range of cloud actors and applications. Four illustrative case studies are presented:

- Infrastructure Optimisation and Network Management for 5G Networks
- Application Optimisation for Network Functions Virtualisation for Virtual Content Distribution Networks
- Application and Infrastructure Optimisation for Edge/Fog computing for Smart Cities
- Capacity Planning for a Big Data Analytics Engine

For each case, we can show that insights and models generated by RECAP can help the stakeholders to better understand their application and infrastructure behaviour. Preliminary results suggests cost savings of more than 25%, up to 20% reduction in bandwidth consumption, and a 4% performance increase.

6.2   Case Study on Infrastructure Optimisation and Network Management—5G Networks

6.2.1   *Introduction*

This case study envisions a system that provides communication services for a variety of industry verticals including eHealth, eCommerce, and automotive. To facilitate the communications of diverse services in

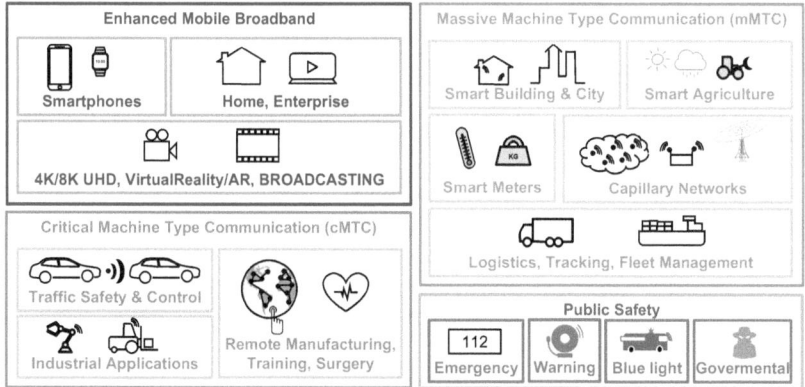

Fig. 6.1 Categories of communication services and example of 5G use cases

different scenarios within the world of fifth generation (5G) networks, the communication system has to support various categories of communication services illustrated in Fig. 6.1.

Each service is needed for a specific type of application serving a particular group of customers/clients. This introduces different sets of characteristics and requirements corresponding to each type of communication service as presented in Table 6.1.

The emergence of 5G mobile networks and the rapid evolution of 5G applications are accelerating the need and criticality of optimised infrastructure as per this case study. Additionally, the management and operation of a 5G infrastructure and network are complex not only due to the diversity of service provisioning and consumer requirements, but also due to the involvement of many stakeholders (including infrastructure service providers, network function/service providers, and content and application service providers). Each of these stakeholders has their respective and different levels of demands and requirements. As a result, a novel solution is required to enable:

- the adoption of various applications under different scenarios on a shared and distributed infrastructure;
- on-demand resource provisioning considering increased network dynamics and complexity; and
- the fulfilment of Quality of Service (QoS) and Quality of Experience (QoE) parameters set and agreed with service consumers.

Table 6.1 Characteristics and requirements of communication services

Service	Characteristics and requirements
Mobile BroadBand (MBB)	Extremely high throughput and user/device mobility
Massive Machine Type Communication (mMTC)	Supports to diverse and massive number of mobile devices, and to enable energy-efficient communications
Mission-critical Machine Type Communication (cMTC)	Ultra-reliable low latency, but high availability and reliability in communications
Public safety (blue light)	Intensely high integrity and availability in services

6.2.2   Issues and Challenges

The communication services between user mobile devices and content services/applications are realised with a set of network functions and numerous physical radio units. In the context of virtualised networks, network functions are virtualised and chained to each other to form a network service providing network and service access to user devices through radio units. A network function virtualisation (NFV) infrastructure is required to accommodate the network function components. Within this infrastructure, virtualised components are deployed in a distributed networking region including the access network, edge network, core network, and remote data centres. Figure 6.2 illustrates a typical forwarding graph of a network service in an LTE network. The network service is composed of multiple virtual network functions (VNFs): eNodeB, Mobility Management Entity (MME), Serving Gateway (SGW), and Packet Data Network Gateway (PGW), and Home Subscriber Server (HSS).

The adoption of such a distributed architecture for the network and its services introduces four major challenges when rolling out a network service:

1. The communication system facilitates various types of applications/services, namely voice/video calls, audio/video streaming, web surfing, and instant messaging. This introduces a high **complexity in understanding individual network services and associated dynamic workloads**.
2. The placement and autoscaling of VNFs are needed by the communication system in order to enable dynamic resource provisioning. VNF components in control and user planes have different features and requirements. As such, to fully address the placement and auto-

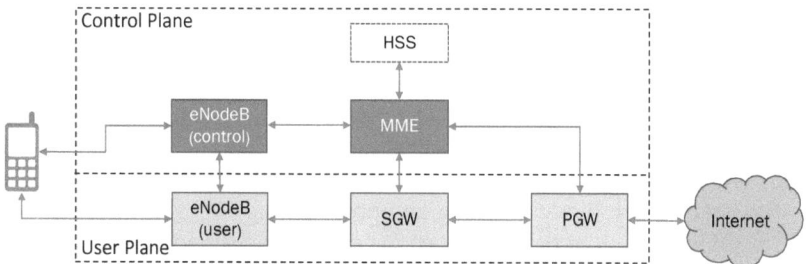

Fig. 6.2 A forwarding graph of a network service in an LTE network

scaling, it is necessary to understand and predict not only the variation of workload and resource utilisation but also the characteristics of the components. **Diversity in requirements and implementation, together with the dispersion of components across the network infrastructure, makes placement and autoscaling of VNFs a significant challenge**.

3. User behaviour needs to be explored for accurate workload predictions. To obtain knowledge of user behaviour, data communicated in control and user planes need to be analysed, and correlations thoroughly investigated. This **analysis is challenging, as one requires domain knowledge regarding** behaviour of **network services and the telecommunication network** more broadly.

4. Multi-tenancy is demanded in emerging 5G mobile networks where multiple network services are deployed and operating on top of a shared infrastructure. Different communication services come with different QoS requirements that desire a capability of adaptation and prioritising in resource allocation and management. In short, **5G brings complexity in the shape of mixed criticality and scale**.

6.2.3   Implementation

6.2.3.1 Requirements

To fully address all the aforementioned challenges, it requires a complete control loop from data collection and analysis to optimisations on both infrastructure and application levels, and further up to the deployment of optimisation plans. This places an overall requirement on a system, such as RECAP, to enable a wide range of automation tasks as listed below:

- Profiling network/service functions and infrastructural resources
- Automated service and infrastructure deployment
- Automated orchestration and optimisation of services and the infrastructural resource planning and provisioning
- Observability of behaviours of the system and services at run-time

To ensure end-to-end QoS (and by inference required service availability and reliability) in a complex large-scale use case, the control loop together with automated solutions need to be capable of resource planning and provisioning in the short- and long term, e.g. in minutes or in months. The solutions are also required to satisfy constraints of multi-tenancy scenarios and multiple network services competing for shared infrastructural resources. In addition, the optimisation aspects of the solution need to manage resource provisioning that achieve utilisation and service performance goals. Moreover, the simulation aspects of the solution need to support evaluation, i.e. impact of changes in planning rules prior to any real deployments.

Table 6.2 summarises the requirements. For each requirement, a set of targeted solutions is presented to illustrate the requirements are met. A simplified mapping is presented but the solution for any single requirement could be derived from one or a combination of multiple solutions listed.

6.2.3.2 *Implementation*

To demonstrate and validate the RECAP approach, a software/hardware testbed in Tieto is used. The testbed, deployed in a lab environment, emulates a real-world telecommunication system to facilitate the development and evaluation of optimisation solutions for end-to-end communications in a 5G network and its applications. Figure 6.3 presents an overview of the testbed in which a distributed software-defined infrastructure is emulated. This is achieved with heterogeneous resources collected from multiple physical infrastructures, located in a wide range of vertical regions, to provide communication and contents services to various applications that form different network services.

From the testbed's perspective, the entire RECAP platform is represented through the RECAP Optimiser, an external component that

Table 6.2 Use case requirements and corresponding RECAP solutions

Requirement	RECAP solution
• Allocation of infrastructural resources to uphold the QoS of a VNF for a provisioning of resource-efficient products • Allocation of a right amount of resources at right locations • Automation of instant capacity checks to support the rollout of new communication services in a timely fashion	• *Workload and workload propagation models* enable estimations of bandwidth and resource utilisation for each VNF, service functions at application and infrastructure levels • The *load translation mapping model* enables quantification of *infrastructural resource utilisations* • *Models for QoS metric assessment* are integrated with the above to fulfil all requirements
• Automation of the optimisation in VNF deployment and autoscaling for service availability and reliability and the minimisation of the overhead and resource utilisation of communication services • Automation of service remediation and infrastructure recovery to uphold required service availability and reliability	• The consolidation of aforementioned models and *optimisation models* facilitates the production of optimisation plans and recommendations for system autoscaling • The *RECAP platform* with automated optimisers empowers the realisation and execution of optimisation plans.
• Predictions of future infrastructural resource demand for resource planning and provisioning in a proactive manner • Detection of resource overbooking for VNFs and service functions to serve optimisation of the resource allocation	• Workload-related models and load translation mapping model enable the predictions of future workload and resource demands • A combination of quantification and predictions of resource demands facilitates overbooking detection
• Support infrastructure and communication service providers to maximise the utilisations of shared infrastructures • Method to prove a VNF is behaving as required on shared infrastructures	• Optimisation plans produced by optimisers with a consideration of scenarios of multi-tenancy and multiple network services enable the maximisation of the utilisation of *shared infrastructural resources* • The aforementioned models combined with telemetry enable VNF and service function performance monitoring and management

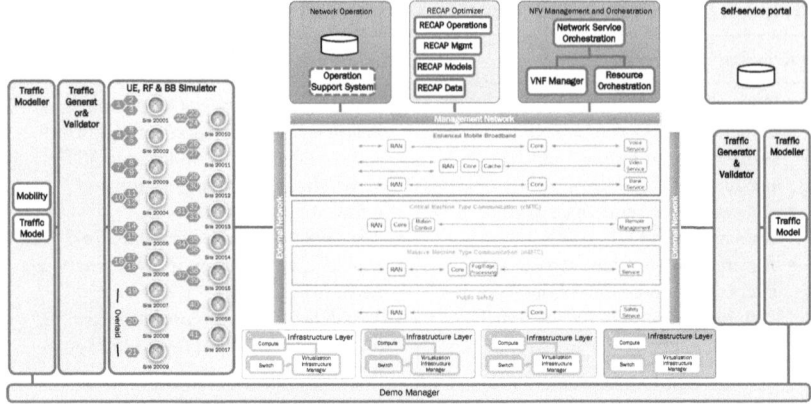

Fig. 6.3   Logical view of the testbed

produces and enacts optimisation plans (through its enactor) on the test-bed. Each plan presents the in-directions of VNF placement and autoscaling across the emulated network infrastructure and is executed by the testbed. Ultimately, results, in terms of both application and infrastructure performance, are collected and evaluated. These results are also fed back to the RECAP platform for further investigation and model improvement.

6.2.3.3 *Deliverables and Validation*
To facilitate validation, multiple validation scenarios covering all the requirements presented in Table 6.2 were defined:

- **Scenario 1**: the placement and autoscaling of VNFs to fulfil QoS constraints required by a given single communication service.
- **Scenario 2**: the placement and autoscaling of VNFs to fulfil QoS constraints required by multiple communication services under multi-tenant circumstances.
- **Scenario 3**: the capability of RECAP simulation and optimisation tools in supporting the offline initial dimensioning the (physical) infrastructure according to traffic demands.
- **Scenario 4**: the capability of RECAP simulation and optimisation tools supporting the offline planning by identifying future (physical) infrastructure needs.

Table 6.3 RECAP deliverables to address validation scenarios

Model/component	Usage
Workload model	To facilitate the implementation of the optimisers as well as optimisation solutions/plans in order to accomplish all the validation scenarios
Application model	
Infrastructure model	
Load translation model	
Infrastructure optimiser	To produce optimisation plans of VNF deployment across the network infrastructure; to directly support validation Scenarios 3 and 4, and together with the application optimiser to address all the scenarios
RECAP optimisation platform	To orchestrate all models and components and enact optimisation plans of VNF deployment that are fed to the testbed

- **Scenario 5**: the observability and fulfilment of given QoS requirements from the VNF level put in the resources provided by the infrastructure.

The relevant models and components that form the solutions to be validated against these scenarios are summarised in Table 6.3.

The application model is developed based on the network services deployed in the Tieto testbed, and workload models are constructed using the synthetic traffic data collected from various experiments carried out within the testbed. The infrastructure network model pertains to the city of Umeå in Sweden but is influenced by BT's national transport network and includes four network tiers (MSAN, Metro, Outer-Core, and Inner-Core). The network topology of the infrastructure is kept symmetrical, without including customisation for real-world aspects for asymmetrical node capacity, for asymmetrical node interconnection, and for asymmetrical link latencies.

6.2.4   Results

This section presents exemplar validation results for the application placement and infrastructure optimisation (Chap. 4). It addresses the problem of VNF placement across the network infrastructure. For the case study, the RECAP Simulator (Chap. 5) was used to calibrate the models used by the Infrastructure Optimiser.

In the experiments, given a network service (Fig. 6.2), the eNodeB is deployed as two separate units on different planes: the Central Unit-User plane (CU-U) and the Central Unit-Control plane (CU-C). Additionally, the SGW and PGW VNF components are located on the user plane and are termed Service Gateway-User plane (SGW-U) and Packet Data Network Gateway-User plane (PGW-U).

The optimisation solutions presented address Scenarios 3 and 4 concerned with placement and infrastructure optimisation. Five placement plans/distributions are identified as the input to the algorithm assuming (1) one forwarding graph per MSAN tier with CU-U as the user-request entry point, and (2) no currently deployed infrastructure. Table 6.4 describes these five placement distributions.

Results obtained are evaluated based on a comparison of provider and customer utility.

In Fig. 6.4, maximum provider vs. customer utility is normalised [0,1] for all distributions. Distributions 1 and 2 only use physical hardware

Table 6.4 Initial placement plans of VNFs

Placement plan	Description
Distribution 1	• CU-U, CU-C, SGW-U, and PGW-U VNFs placed on the MSAN resource sites • MME/SGW-C/PGW-C VNF placed on the Outer-Core resource sites
Distribution 2	• CU-U, CU-C, SGW-U, and PGW-U VNFs placed on the Metro resource sites • MME/SGW-C/PGW-C VNF placed on the Outer-Core resource sites
Distribution 3	• CU-U and CU-C VNFs placed on the MSAN resource sites • SGW-U and PGW-U VNFs placed on the Metro resource sites • MME/SGW-C/PGW-C VNF placed on the Outer-Core resource sites
Distribution 4	• CU-U and CU-C VNFs placed on the MSAN resource sites • SGW-U, PGW-U, and the MME/SGW-C/PGW-C VNFs placed on the Outer-Core resource sites
Distribution 5	• CU-U VNF placed on the MSAN resource sites • CU-C and SGW-U VNFs placed on the Metro resource sites • PGW-U and the MME/SGW-C/PGW-C VNFs placed on the Outer-Core resource sites

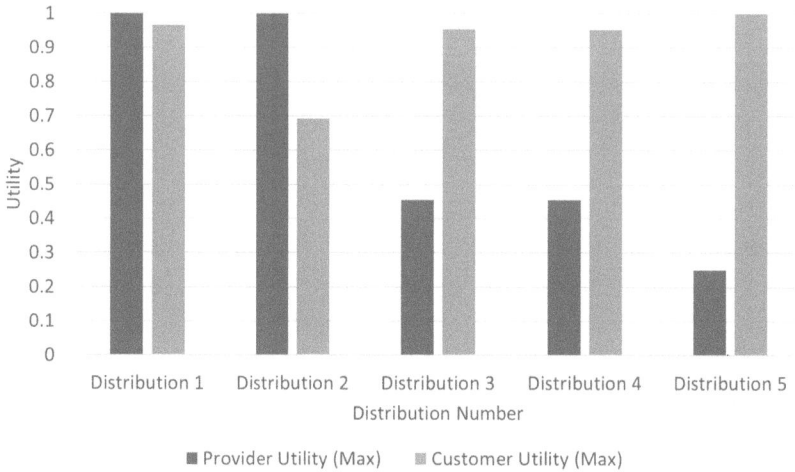

Fig. 6.4 Maximum provider and customer utility of each distribution

across two tiers (as per Table 6.4) and hence exhibit the highest provider utility; this decreases by approximately 50% as the distributions start to include more than two tiers in the placement. This is understandable as more infrastructure needs to be deployed and maintained. Customer utility is highest for Distribution 5, which has three tiers MSAN, Metro, and Outer-Core (as per Table 6.4) included in the distribution. The lowest utility is when no VNFs are placed at the edge, i.e. Distribution 2, which has no MSAN; this is to be expected as the end-user request faces larger processing latency in travelling further into the network.

Figure 6.4 maps the normalised provider utility and normalised customer utility of each VNF placement. The figure shows that a provider can manage its deployments by fixing the provider utility or customer utility in a way that balances business considerations.

For example, in Fig. 6.5 the provider utility is centred on 50% to ensure customer utility is centred on 75%. The intersection of threshold lines (the highlighted section in grey) identifies a set of placements that are optimal for each individual forwarding graph of the use case while satisfying defined constraints including application and infrastructure provider perspectives. The provider could choose Distribution 1, 2 or 4. However, Distribution 2 has poorer customer utility (no MSAN, higher latency) and so is disregarded. Distributions 1 and 4 utilise MSAN and Outer-Core

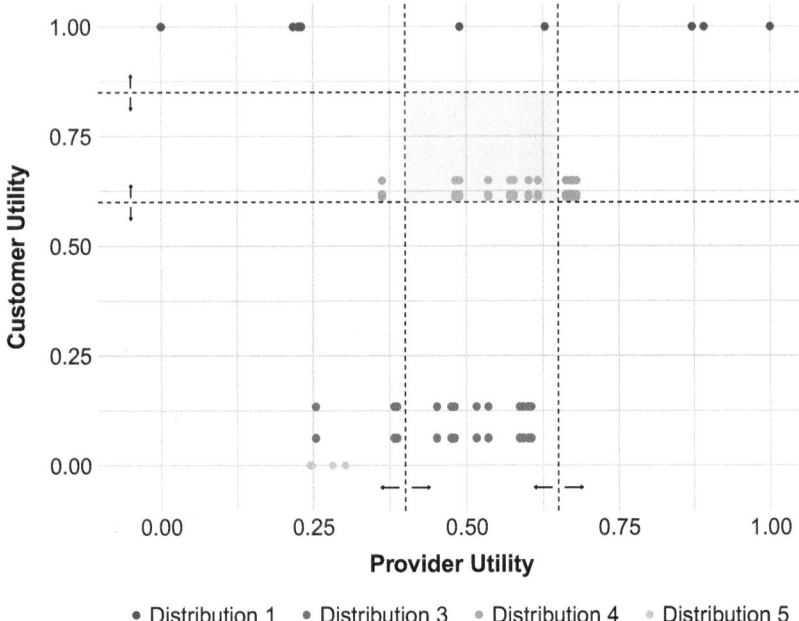

Fig. 6.5   Provider utility vs. customer utility for different distributions

infrastructure and have comparable customer utility. However, Distribution 1 has the higher provider utility, and thus would be the best option with caveats. It would be the best option if "consolidation" was the most important factor to the business, but not necessarily the best option if "flexibility" to service future requests was more important. In the latter case Distribution 4 is a better option because one's current customer is satisfied (compared to Distribution 1) but the provider has significant up-swing capacity.

Figure 6.6 below illustrates simulation results for the same distributions without Distribution 2 which was disregarded due to no MSAN capacity. The graph represents all infrastructure and all remaining distributions. Utility is combined for simplicity (y-axis) and is graphed against 3 scenarios (x-axis) "normal day", "event", and "24% growth". It should be more apparent that for the same scenarios and constraints as above, Distributions 1 and 4 remain the best options. Distribution 1 remains the best option and for this simulation exercise could cope with the defined *event* and *growth* scenarios. But what is a little less obvious is that its utility remained essentially static while

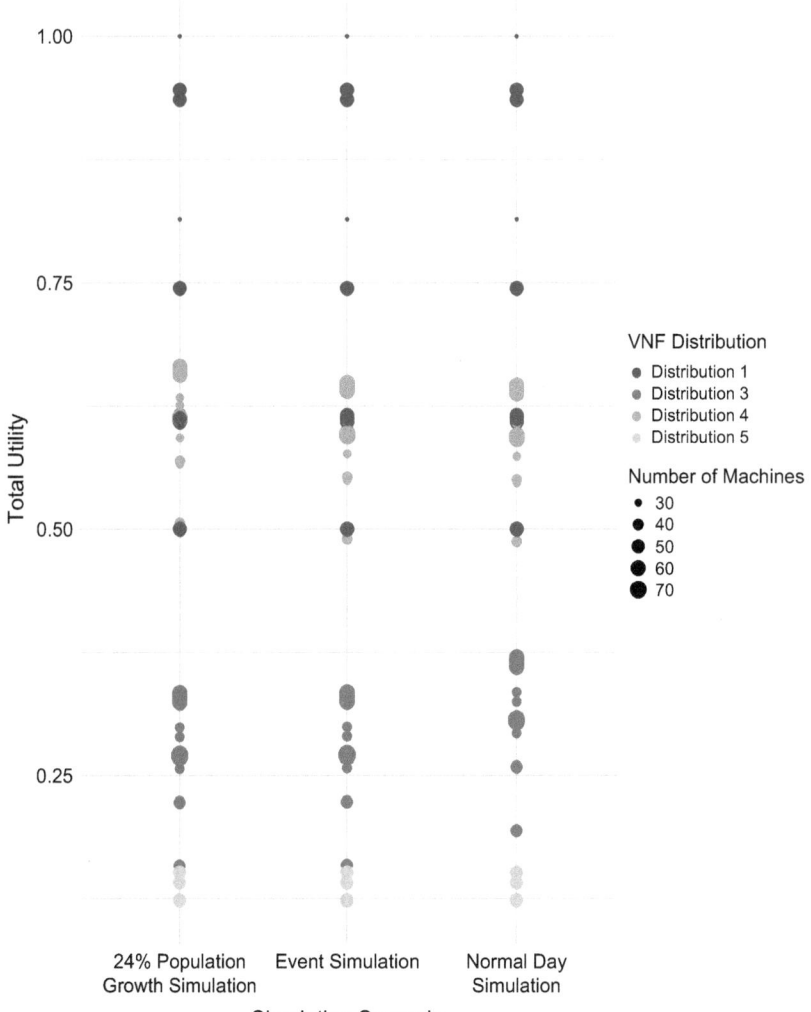

Fig. 6.6   Total utility for normal day, event, and 24% growth scenarios

the utility for Distribution 4 starts to trend upwards from the *normal day* to *event* to *24% growth* scenario. This is primarily driven by provider utility improving as utilisation of physical assets improve; this scenario offers considerably greater capacity for future *growth/event* scenarios.

6.2.5   Summary

For this case study, an extensive evaluation and validation were performed for the scenarios outlined and utilising the models, components, and technologies described in Chaps. 1, 2, 3, 4, and 5. Current results contribute to more effective decision making for infrastructural resource dimensioning and planning for future 5G communication systems. The value presented is supporting informed, automated (if desirable) decisions similar to the consolidation vs. flexibility example illustrated in the Distribution 1 versus Distribution 4 options above.

6.3   Case Study in Network Functions Virtualisation: Virtual Content Distribution Networks

6.3.1   Introduction

Network Function Virtualisation (NFV) replaces physical network appliances with software running on servers. Content Distribution Networks (CDNs) offer a service to content providers that puts content on caches closer to the content consumers or end users. Traditional Content Distribution Network (CDN) operators install customised hardware caches across the globe—sometimes within an Internet Service Provider's network and sometimes in third-party co-located data centre facilities. Each CDN operator develops its own caching software with unique features, e.g. transcoding methods, management methods, and high availability solutions.

As of now, network operators such as BT have hardware from each of several CDN operators deployed at strategic points in its network. However, this creates several potential issues:

- it is hard to organise sufficient physical space (in "telephone exchange" or "central office" buildings for instance) to support all the CDN operators;
- a lot of energy is needed to power and cool all the equipment; and
- a lot of physical effort is needed when a new CDN operator arrives or an existing one disappears.

Such factors make it attractive to consider a Virtual CDN (vCDN) approach that aims to replace the multiple customised physical caches with standard servers and storage running multiple virtual applications per CDN operator. This lowers CDN and network operator costs and allows the content caches to be put closer to the consumer, which improves customer experience. Also, the barriers to entry for new virtual CDN operators are likely to be lower than for a physical CDN operator.

6.3.2 Overview and Business Setting

Broadband traffic on BT's network of 50% of broadband traffic on BT's network originates from the content caches operated by the CDN operators. At the time of writing, BT hosts CDN operators customised cache hardware in two to eleven compute sites in the UK to reduce the amount and cost of Internet peering traffic. If the caches were installed in BT's thousand edge nodes (also known as Tier-1 MSANs (Multi-Service Access Nodes) or "Telephone Exchanges"), the cost of delivering content would be reduced by 75% and BT would reduce its network load significantly. However, the CDN operators are unlikely to want to install their hardware at up to a thousand locations in the UK; for some international CDN operators, a single compute site in London is sufficient for the entire UK.

The vCDN proposition is that BT could install the compute infrastructure at its edge sites and offer an Infrastructure-as-a-Service (IaaS) offering tailored towards CDN operators. The CDN operators would install and manage their own software on BT IaaS and thus they would maintain their unique selling points and ownership of the content provider customers. This is a potential win-win scenario: the network and CDN operators reduce operating costs and consumers get better service (Table 6.5). There are, however, several technical challenges to designing and operating a vCDN service, not least performance, orchestration, optimisation, monitoring, and remediation. These are discussed later in Sect. 6.3.3.

An abstract representation of the BT UK network topology is shown in Fig. 6.7. The real locations of BT network sites are shown in Fig. 6.8; the black dots represent BT's 5600 local exchanges, of which c. 1000 are MSANs and c. 100 BT's Metro sites (Ofcom 2016). The 4500 local exchanges that are not MSANs are considered unsuitable for deployment of caches as they do not contain enumerates the most important vCDN use case requirements.

Table 6.5 vCDN use case requirements and corresponding RECAP components

Requirement	RECAP solution
• Optimise cost of compute and storage infrastructure vs. cost of network bandwidth.	• CDN traffic forecast using a Seasonal Autoregressive Integrated Moving Average (SARIMA) model enables accurate prediction of demand. • CDN load translation models to calculate how much compute resource is required for the forecast workload. • Cache placement optimisation using state-of-the-art AI and genetic algorithms. • CDN application model to calculate where caches should be placed in the BT topology. • Infrastructure model of BT's network and compute infrastructure. • Simulator to calculate utility of infrastructure placement options.
• Take account of uneven distribution of consumers and traffic.	• The RECAP methodology addresses this (further work is required to add customer distribution into the optimisations). • The application model takes account of BT's topology.
• Different CDN operators may have different optimal locations for their caches	• Application model can be run per CDN operator.
• Network operator must take into account the potential demands across multiple CDN operators	• Infrastructure model can aggregate demand from the application model.
• Content traffic has a 2:1 peak-to-mean ratio and is highly seasonal with daily, weekly, and annual patterns. Power could be saved by turning off infrastructure when not required. • CDN operators will need tools to support a near real-time decision to activate or deactivate their caches.	• Application cache placement optimiser is dynamic and can adjust according to traffic load.

(continued)

Table 6.5   (continued)

Requirement	RECAP solution
• Total bandwidth consumed by content is consistently increasing (c. 50% per annum); therefore, the network operator needs to constantly invest in adding more transmission or more vCDN infrastructure nodes and capacity to the network. Network operators need to improve the accuracy of future traffic predictions and where investments should be made so that infrastructure gets installed just in time and customer experience is always good.	• The collection of RECAP workload predictors, models, and optimisers find solutions, which optimise cost and performance according to network operator preferences.

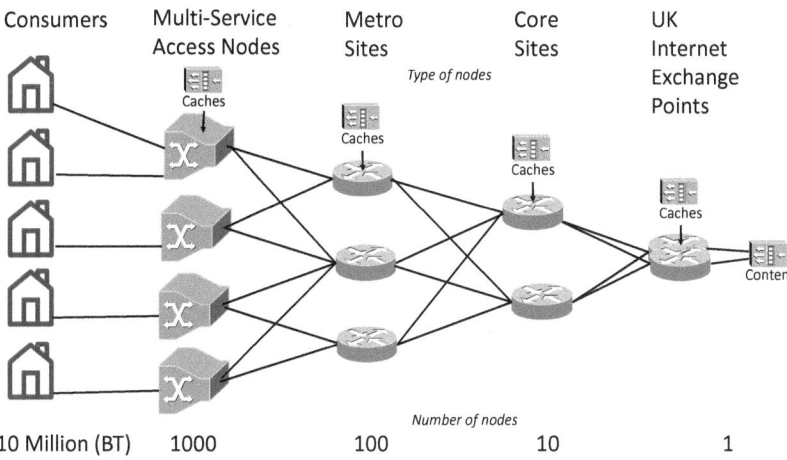

Fig. 6.7   Abstract representation of BT UK network topology

Fig. 6.8   BT network locations in UK

6.3.3   Technical Challenges

The vCDN technical challenges can be grouped into several areas as shown in Table 6.6 below.

Table 6.6   vCDN technical challenge and RECAP solution

Technical challenge	RECAP solution
Performance of virtualisation technologies, especially virtualised storage.	BT is conducting experiments to verify and improve the performance and orchestration of CDN virtualisation technologies.
Orchestration of a multi-tenant vCDN service and infrastructure.	BT is building an orchestration system proof of concept using OSM and OpenStack for its vCDN service.
• Optimisation of placement and scaling of vCDN system. • BT decides where to install and grow or reduce the infrastructure. • The CDN operators decide where to install, grow, and reduce their virtual machines or containers for their caches. • BT will have to optimise how much of its infrastructure it dedicates to CDN operators. • CDN operators will have to optimise how much resource they need to consume. • Each CDN operator will be independent and may experience different traffic loads. • Installing infrastructure requires the planning of the provisioning of hardware many months in advance, although once installed servers may be turned on and off, to reduce power consumption, in ~15 minute time periods. • CDN operators may activate and deactivate their cache virtual machines or containers very quickly in the 5 minutes to sub-second timeframes and hence need more real-time optimisation than the infrastructure. • CDN operators may choose to deactivate their software to reduce infrastructure rental charges.	RECAP methodology can automate the decision making for the optimisation, scaling, monitoring, and repair of vCDN systems using modelling and statistical techniques. • RECAP methodology is built around "separation of concerns" addressing the need for network and CDN operators to be treated separately. • RECAP forecast model can enable decisions to be made and acted upon just in time to optimise power consumption. • RECAP application model adjusts to traffic dynamics.

(*continued*)

Table 6.6   (continued)

Technical challenge	RECAP solution
• Monitoring and repair of the vCDN system. • Each CDN operator will have proprietary methods for the monitoring and remediation of their CDN software. • Many operate an architecture that is fault tolerant, with a caching hierarchy, where loss of a leaf will result in content being served from a cache higher in the hierarchy. • CDN operators also have advanced load balancing mechanisms where a consumer's initial request is switched to the best cache, according to load and location, and content is "chunked" and distributed so any failure mid-session will be recovered from.	• The RECAP methodology can automate the decision making for the optimisation, scaling, monitoring, and repair of vCDN systems using modelling and statistical techniques • The CDN operators' architectures permit re-optimisation of CDN cache locations and scale with minimal impact on service.
• Detection and mitigation of impact of "noisy neighbours".	• Proprietary solutions exist to monitor the quality of the content delivered to consumers. • Further work is required to feedback these quality measurements into a CDN cache placement optimisation and orchestration solution.

6.3.4   Validation and Impact

The RECAP consortium is engaged with various CDN operators to develop fine-grained infrastructure and application models to develop optimisation strategies for Virtual Content Distribution Networks (vCDN). The resulting strategies will aid BT to improve the accuracy of their planning and forecasting, reducing infrastructure investment while still giving BT's customers a superior web browsing and video streaming experience. RECAP methods will reduce the amount of human support BT's vCDN planning process requires, enabling BT to be more agile and cost efficient (Fig. 6.9).

Preliminary experimentation results are promising regarding the utility of RECAP for BT. They suggest:

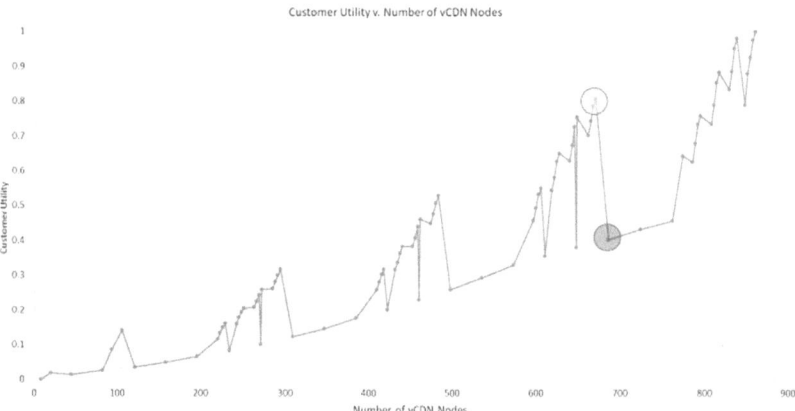

Fig. 6.9 Customer utility vs. number of vCDN sites

1. The SARIMA models provide one-hour ahead workload forecasts of 11.5% accuracy with 90% confidence. This should be sufficient for CDN operators to pre-emptively adjust the sizes and number of caches running while infrastructure operators should be able to shut down or power up servers to minimise power wastage.

2. The RECAP DTS framework demonstrated the value of a caching hierarchy when compared to a single layer of caches at the MSAN. The results suggest a caching hierarchy can improve Provider Utility by up to 24% (see Fig. 6.10) and doubles Customer Utility at intermediate stages of infrastructure deployment, as illustrated in Fig. 6.9. Further, when compared to BT's own optimisation strategy, it improved Provider Utility by up to 6.4% at certain intermediate stages of infrastructure deployment. The BT and RECAP optimisation both converged on deploying the maximum number of 860 nodes because the caching business case is very compelling, i.e. deploying a cache at a site always saves money and improves customer utility. For a use case where the business case is more marginal, the RECAP methodology can find a solution more significantly optimal.

3. The RECAP Application Optimisation (autoscaling and simulation) systems provide BT with the ability to both improve on baseline cache deployment scenarios (through comparative analysis of alternative deployment scenarios) and in run-time adapt to unfore-

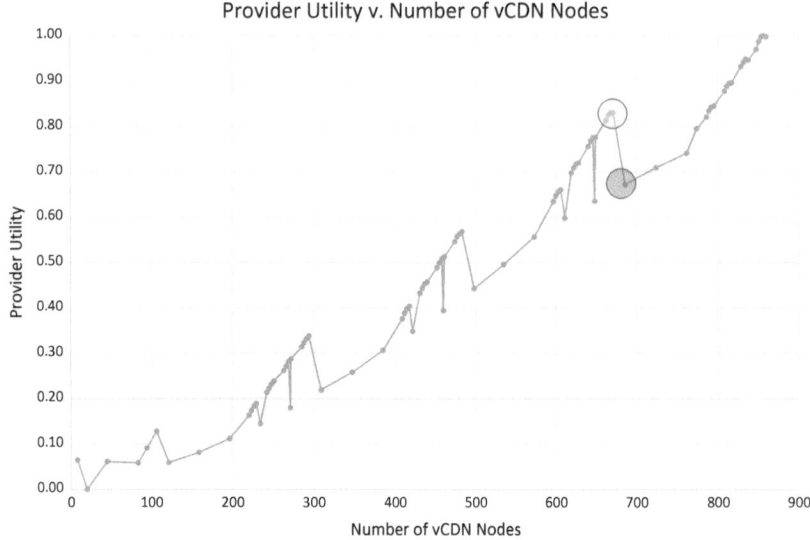

Fig. 6.10   Provider utility vs. number of vCDN nodes

seen and unexpected changes in workload. Simulation allows experimentation with alternative deployment strategies and evaluates the impact of changes in infrastructure, as well as application topologies and caching strategies. Validation experiments demonstrate a 4% improvement in cache efficiency when serving realistic workloads as well as the ability to efficiently adapt the amount of cache capacity deployed in heterogeneous and hierarchical networks to changes in request and network traffic patterns.

These results suggest that by implementing RECAP, BT and CDN operators can benefit from both decreased cost and increased competitiveness through:

1. Providing more accurate modelling, infrastructure dimensioning, and resource allocation across the chain of service provision to support better infrastructure planning.
2. Rapid accurate autoscaling to support fluctuations in demand and avoid under and over booking of resources.

3. Leveraging existing infrastructure and avoiding additional capital expenditure.
4. Reducing staffing requirements and freeing up valuable IT expertise.
5. Increased revenue through:
   (a) Delivering and maintaining high QoS.
   (b) Shortening the time for CDN operators to access infrastructure and accelerate revenue generation.
   (c) Reducing time to market for infrastructure and applications deployment.

6.4   CASE STUDY IN EDGE/FOG COMPUTING FOR SMART CITIES

6.4.1   Introduction

This case study integrates RECAP mechanisms related to resource reallocation and optimisation in a proprietary distributed IoT platform for smart cities, called SAT-IoT. Hence, it demonstrates the capabilities to integrate RECAP components into third-party systems as well as the immediate benefits of using the RECAP approach for optimisation.

Being IoT-centric, this case study deals with hardware-software infrastructures and vertical applications as well as mobile entities and devices that move over the area of city. It is built on the assumption that smart cities provide infrastructure for handling IoT network traffic in a zone-based manner as shown in Fig. 6.11; wireless networks complement wired networks to form a hybrid network (Sauter et al. 2009); and these hybrid networks include the cloud nodes, edge nodes (IoT gateways), and further mid (fog) nodes. Mid nodes are connected to the cloud and to each other forming a mesh network. Edge nodes receive data from wireless devices located in the same geographical area. Groups of edge nodes are connected to a mid node. Edge nodes are usually not connected to each other. In this kind of scenario, it is necessary to manage the IoT network topology to adapt to moving users and changing data streams. Such a topology administration will facilitate the dynamic deployment of distributed IoT applications, the interconnection of devices in the IoT platform, and the data exchange among platform network nodes.

Consequently, this use case study requires:

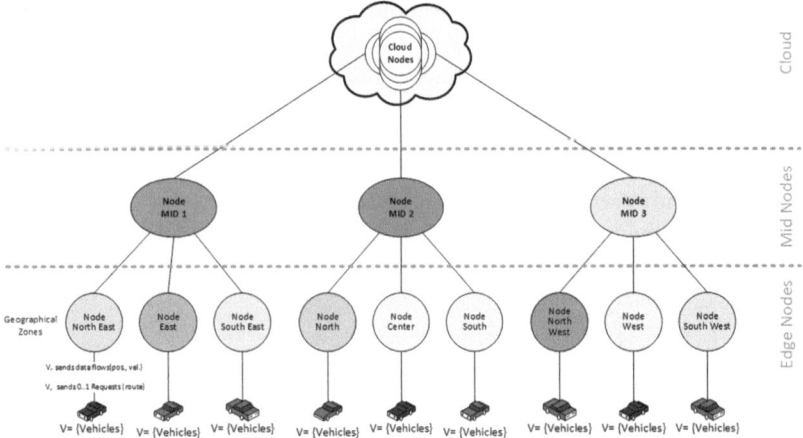

Fig. 6.11   Example of IoT hybrid network for mobile devices

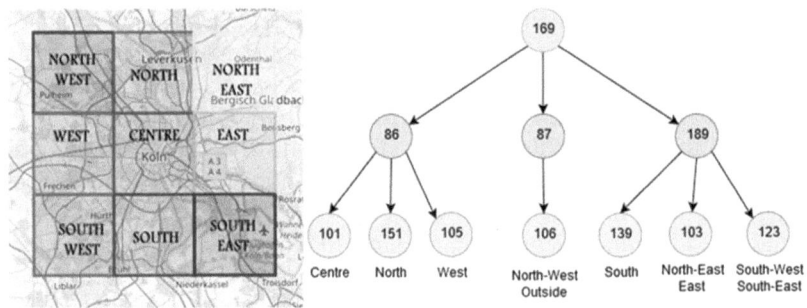

Fig. 6.12   Smart city structure

- The capability to dynamically optimise the applications' communication topologies under an "Edge/Cloud Computing Location Transparency" model. In particular, it requires the optimal reallocation of data flows periodically at run-time to reduce bandwidth consumption and application latencies.
- A mechanism to consider limitations of the physical topology when planning the virtual topology.

For the sake of demonstration and validation purposes, the SAT-IoT platform in RECAP is running a distributed Route Planning and City Traffic Monitoring application. Figure 6.12 illustrates this scenario based on the city of Cologne. The city area is distributed into nine geographical regions, each of which has its own edge compute node.

6.4.2 Issues and Challenges

The formulated scenario presents some challenges with respect to the current cloud IoT/smart city systems, the edge/fog computing paradigm, and the ICT architecture optimisation. The issues and challenges addressed in this case study include:

- Adoption of edge computing models to process the massive data generated from different IoT devices at their zone edge nodes in order to improve performance.
- Integration of edge, fog, and cloud paradigms to develop dynamically configurable IoT systems, to achieve better optimisation results for both applications and underlying infrastructure.
- Dynamic management of smart city environments based on distribution of mobile devices and users and their resource demands.

6.4.3 Implementation

6.4.3.1 Requirements

In order to improve the performance of the IoT system, the edge computing model seeks to process the massive data generated from different IoT devices at their zone edge nodes. Only the processing results are transmitted to the cloud infrastructure or to the IoT devices, reducing the bandwidth consumption, the response latency, and/or the storage needed (Ai et al. 2018).

Considering an IoT system that uses a hybrid network similar to Fig. 6.11, any application that processes data from zones, North and Centre, cannot naively run data processing in the zones since it requires information from both zones. A conventional cloud computing architecture is not well suited to applications where the location of devices changes, where the volume of data received in each edge node varies dynamically, or where the processing needs data from different geographical areas. Car route planners and city traffic analysers are good examples of smart city applications that make calculations with the information received from connected cars located in different zones of the city.

Thus, to process the data from, for example, the North and Centre zones, it is necessary to send it to a processing node. As these edge nodes are not connected to each other, the data from one or both zones need to pass through Mid86. Indeed, Mid86 would be the closest node to run processing for both affected zones. When considering the scenario on a larger scale with multiple zones and applications that require data from all zones to complete the processing,[1] the task of finding the most suitable processing node is non-trivial. Furthermore, where there are significant constraints on deployment capacity constraint, the underlying infrastructure becomes a factor too. Consequently, an IoT platform for smart city applications must be able to:

- integrate of cloud, fog, and edge computing models;
- manage the smart city data network topology at run-time;
- use optimisation techniques that support processing aggregated data by geographical zones; and
- monitor the IoT system and the optimisation process in run-time.

6.4.3.2   The SAT-IoT Platform

The IoT platform which forms the basis of this case study is based on the SAT-IoT platform. Its core architectural concept is edge/cloud computing location transparency. This computational property allows data to be shared between different zones and to be processed at any of the edge nodes, mid nodes, or cloud nodes.

The concept of edge/cloud computing location transparency is realised by two of the entities in the SaT-IoT architecture, the IoT Data Flow Dynamic Routing Entity, and the Topology Management Entity (see Fig. 6.13). They support a cloud/fog programming model with the capability of managing the network topology at run-time while also providing the necessary monitoring capabilities to understand the usage pattern and capacity limitations of the infrastructure. While they provide the necessary capabilities to reconfigure the topologies and data flow, they lack the capability to derive the best-possible placement of the data processing logic. This is realised by integrating the RECAP Application Optimiser in the SAT-IoT platform.

---

[1] SAT-IoT is capable of managing and supporting multiple applications over the same IoT data network.

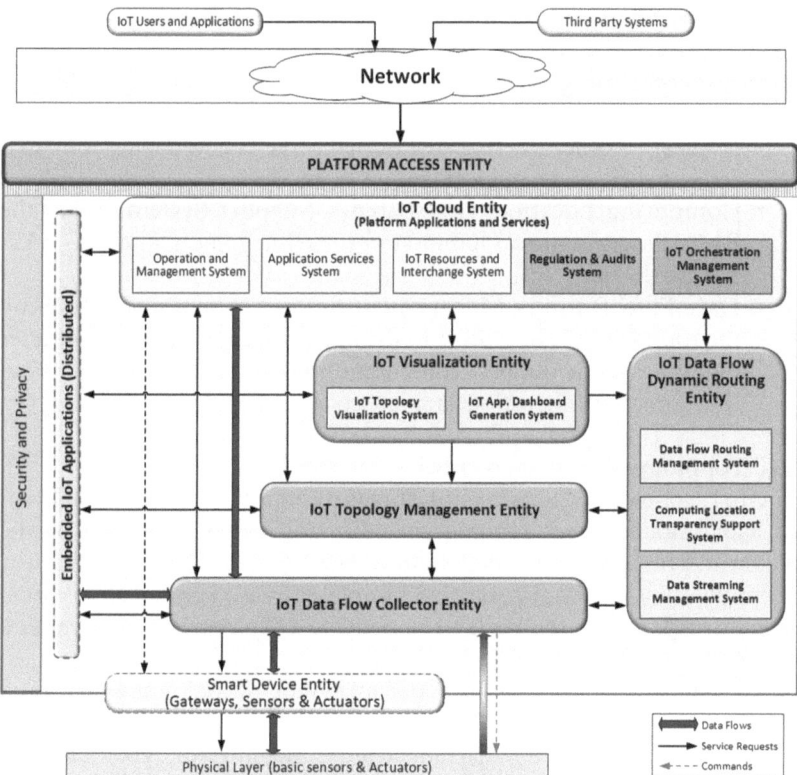

Fig. 6.13 SAT-IoT platform architectural model

IoT Topology Management Entity

IoT Data Flow Dynamic Routing is the cornerstone of SAT-IoT. It dynamically manages IoT data flows between processing nodes (cloud nodes, edge nodes, and smart devices). In addition, this entity includes a distributed temporary data storage system to support data streaming and local processing services. In this case study, data flows are both the sets of data sent by the cars (position, speed, fuel consumption, etc.) and the route calculation requests between two locations in the city.

The IoT Data Flow Dynamic Routing Entity comprises:

- **Data Streaming Management System**: Provides the mechanisms to transfer IoT data flows directly from nodes (e.g. edge nodes or smart devices) to other internal or external services and applications that request them (on a publish/subscribe model).
- **Computing Location Transparency Support System**: Wraps the RECAP Application Optimiser for its integration into the SAT-IoT platform.
- **Data Flow Routing Management System**: Responsible for setting the routing of the data flows to the optimum computation node after inquiring about the best computation node for the data flow from the Computing Location Transparency Support System.

IoT Topology Management Entity

The IoT Topology Management Entity is responsible for the definition of an application network topology in every IoT system deployed by the platform. This application network topology defines which SAT-IoT entity communicates with which other SAT-IoT entities. The communication structure is based on the available underlying IT infrastructure (computation nodes and data network).

The application topology is defined as a graph of computing nodes and links between them, and it includes a variety of attributes like node features (CPU, Memory, etc.), data link features (bandwidth), geolocation of the node, use of resources (hardware and communication metrics), etc. With this definition, the system dynamically manages the physical hardware topologies, enables updating the logical structure of the topologies at any time, and includes a monitoring system that continuously provides the status of nodes and links in terms of performance metrics (consumption of CPU, memory, storage, bandwidth, etc., and also data flows crossing the network).

The three main functions of the **Topology Management Entity** are:

- **IoT Topology Definition**: It enables the modelling of the IoT architecture as an enhanced graph in which nodes are the hardware elements with processing capabilities. The nodes in the graph are defined containing all their attributes (node type, CPU, RAM, location, etc.). Edges in the graph correspond to data links and have their attributes as well.

- **Topology Management**: It is a set of services to query and modify the IoT topology definition to maintain the consistency between the physical installations and their definition in the platform. It supports the model of Edge/Cloud Computing Location Transparency supported by the platform.
- **Topology Monitoring**: It continuously gathers and stores metrics of each node and edge. It also provides these metrics to other internal systems (IoT Topology Visualisation System or Embedded Applications) and external systems (third-party applications and systems).

Application Optimisation

To find the optimal location of the data processing logic, the optimiser needs to consider response latency, bandwidth consumption, storage, and other properties. Furthermore, the selection of the computation nodes might change dynamically as the conditions of the system may vary over the time (shared data, application requests, data volume, network disruptions, or any other relevant issue).

6.4.3.3 Implementation

For realising optimisation support, SAT-IoT integrates a RECAP application optimisation algorithm. Using this algorithm, SAT-IoT can decide, in real time, the optimum node of the IoT data network to process a given data flow. The integration of the application optimisation algorithm is implemented in the Computing Location Transparency Support System module, part of the IoT Data Flow Dynamic Routing Entity (see Fig. 6.14).

The application optimisation algorithm uses IT resource optimisation techniques, graph theory (based on the topology graph definition), and machine learning processes to predict the needs of the system in the short term. The prediction considers the current state of the systems, e.g. metrics, IT resources used, links bandwidth consumption, application latencies, distribution of nodes across the topology, and the data flows involved in each node. A RECAP **Non-Dominated Sorting Genetic Algorithm** is used to calculate the optimum node to process the application data flows received in the last time period.

The Application Optimiser systematically receives the virtual topology and the data flows (route calculation requests and information about them) for the last time period. It then calculates a cost function to move the flows to each server and finally selects the node with the minimum

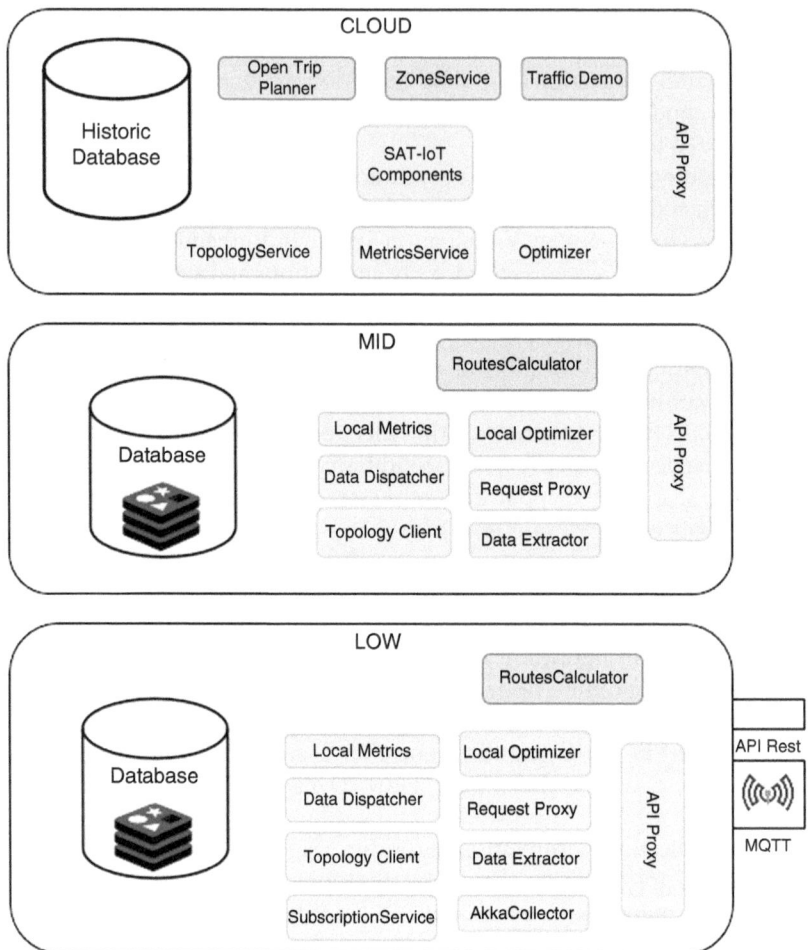

Fig. 6.14   SAT IoT platform high-level conceptual architecture

value of the cost function. The application receives the node selected and requests the platform to configure the data flow routing to send the data flows to the optimum computation node.

The city traffic monitoring application makes use of the optimisation service provided by SAT-IoT and provides a user interface to execute the optimisation on demand. In a production setting, the Application Optimiser would be run automatically based on intervals. Every time the optimisation is executed, SAT-IoT automatically changes the virtual network routing configuration in order to send data flows to the optimum node for processing.

6.4.4   Validation Results

To validate the system, SAT-IoT runs a distributed Route Planning and City Traffic Monitoring application using a dataset of Cologne city traffic edited by the Simulation of Urban Mobility (SUMO) Eclipse project as an input. Software entities emulate cars moving in the city and send data such as position, velocity, and road conditions captured from sensors periodically.

The IoT platform and application have been deployed in a virtual infrastructure with a topology as shown in Fig. 6.15 where seven edge servers associated to ten areas of Cologne city are used (nine city zones and an additional zone to cover traffic close to those defined zones). Three mid nodes have been deployed to group sets of three areas, and a group of virtual servers acts as the cloud infrastructure. Runs cover traffic simulation of two hours.

As discussed, SAT-IoT makes use of the RECAP Application Optimiser (see Fig. 6.13). Periodic optimisation is switched off and manual optimisation enabled to allow the user to execute the RECAP Application Optimiser on demand while the simulation scenario is running. This allows optimisations to be performed at different execution times where the conditions and status of the platform may vary. For instance, route requests and operational vehicles vary over time and executing the Application Optimiser at different points in time results in different optimisation results as shown at the top of Fig. 6.15. Here, the orange circle in the upper left diagram represents the optimal node for data processing at that point in time. Similarly, the chart at the upper right side shows the cost function for the selected node compared to cloud-based data processing.

In the lower left chart, Fig. 6.15 shows the values and results obtained from the optimisation process executed nine times during an experiment. In the first optimisation, the optimum destination node changes from the cloud to Node 86. The line chart on the right shows an immediate cost reduction. The table on the bottom left shows the cost saving/additional cost of moving the data to different nodes. In this case, the optimum node shows a reduction of 3311 cost points compared to moving the data to the cloud node. These results are evident in the bar chart at bottom right too.

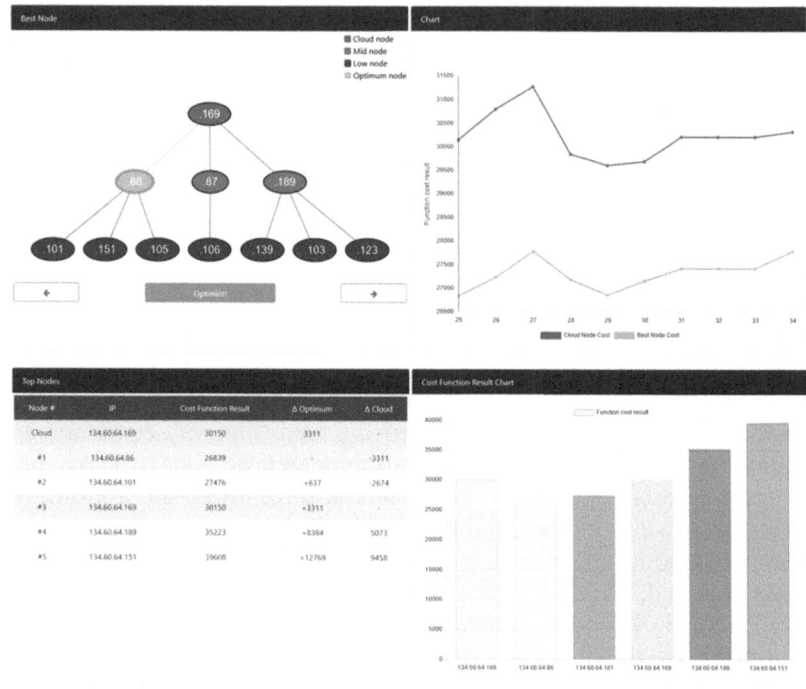

Fig. 6.15 Optimisation results

Figure 6.16 plots the overall amount of data transferred per time unit. It compares a standard cloud-based processing approach with an approach using RECAP-Optimised message routing. As can be seen, it shows a significant reduction of transferred data and hence bandwidth consumption.

6.4.5   Results

Implementation of the RECAP Application Optimiser (1) reduced bandwidth usage by up to 80% compared to a cloud-only processing of data, and (2) reduced the overall latency and improved the user experience by reducing the overall number of hops to send the data flows to the optimum processing node by up to 20%.

In summary, the benefits of the implementation of the RECAP Application Optimiser embedded in the SAT-IoT platform include (1)

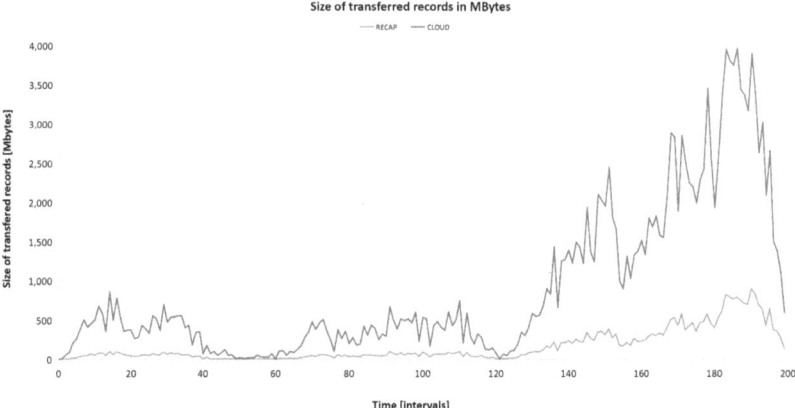

Fig. 6.16 Number of records transferred for SAT-IoT running route planning and city traffic monitoring application using cloud-based processing and RECAP-optimised processing

automated continuous optimisation; (2) enabling dynamic changes to computation nodes in an IoT network topology without administrator intervention to fulfil the efficiency criteria defined for the IoT system; and (3) global transparency across the entire IoT system, which together result in significantly reduced costs and increased quality of service.

6.5   CASE STUDY IN CAPACITY PLANNING: BIG DATA ANALYTICS SEARCH ENGINE

6.5.1   Introduction

This case study illustrates how components of the RECAP approach, namely data analytics as well as simulation and planning, can be used for capacity planning for and small-to-medium-sized enterprises.

Linknovate (LKN) is a Spanish SME that develops and markets a cloud-base data analytics and competitive intelligence platform and service. LKN's primary market is in the US. LKN generates knowledge insights by aggregating large amounts of (heterogeneous) research and scientific data using data mining and data analytics techniques for their clients. At the time of writing, LKN had indexed over 20 million documents, over 30

Table 6.7   LKN platform components

Nodes	Components
Online—index and search nodes	ElasticSearch v5 providing the search functionality and storing processed and structured data: • 9 data nodes (Azure DS12 v2) running Debian GNU/Linux with 4 processing cores and 28GB of RAM, where 24 shards of data with replicas of old and fresh data are stored and processed • 1 client node (Azure DS14 v2) running Debian GNU/Linux with 16 processing cores and 112GB of RAM, which coordinates and aggregates the search results.
Online—web processing node	Azure D12 v2 running Debian GNU/Linux with four processing cores and 28GB of RAM. Service running: Nginx server with Django-based web app. This node serves as the LKN web platform.
Online— database node	Azure DS4 v2 virtualised machine) running Debian GNU/Linux with 8 cores and 28 GB memory running Postgresql, Cassandra and MongoDB. This node stores diverse user information, mailing management, and storing input form data from users.

million expert profiles, over 2 million entity profiles, and more than 200 million innovation topics.

LKN manages vast amounts of information through different offline and online layers. The offline layer, Data Acquisition, comprises several pre-processing components working in parallel over raw data to homogenise structure and identify entities and semantic relations. The online layer, Processing and Indexing, is done over a virtual cluster of search nodes based on ElasticSearch (ES). Finally, the Web and Search layer is where user queries execute several internal queries over LKN indices, retrieving the data to be displayed in the User Interface (UI). User queries are received by the virtual Nginx web server that also renders the results pages. The LKN platform is deployed on a heterogeneous technology stack on the Microsoft Azure cloud with three types of nodes: web processing nodes, database nodes, and the aforementioned index and search nodes. An overview of the LKN platform components is provided in Table 6.7.

6.5.2   Issues and Challenges

Small businesses typically operate in constrained business environments with a tension between scaling for growth and cashflow. While cloud computing provides significant benefits in terms of cashflow management and scalability, controlling consumption and managing complex cloud infrastructure with a small IT team are significant challenges. Small businesses may not be able to accommodate reactive approaches to infrastructure provisioning (given the elevated warm-up times) and could save costs and improve QoS by using predictive solutions. Such solutions should allow effective and efficient provisioning/deprovisioning of cloud capacity by predicting spikes in demand in the short- and medium term and enabling boot-up instances in advance thereby addressing consume pattern prediction by geographic region and accurately anticipating periodic time-based traffic patterns.

In this case study, LKN overprovision nodes in Azure to cope with unexpected or irregular request peaks by users with a focus on serving the Eastern US market. LKN would like to optimise their cloud resources to reduce the cost of overprovisioning and avoid platform replication in non-core geographic markets. RECAP Data Analytics and Simulation and Planning methodologies and tools were used to support LKN in the capacity planning.

6.5.3   Implementation

6.5.3.1   RECAP Data Analytics

Step 1: Exploratory Data Analysis

Web Server Error Analysis
Firstly, LKN data were evaluated from a quality perspective. Given that the workload is based on the number of user queries, the errors were evaluated (number of invalid requests) reported by the web server as per Table 6.8 below.

Although 78.78% of the queries were successfully answered by the search engine, the number of errors is very high for this kind of service.

Request Source Analysis
In a second step, the source of all requests was analysed and ordered by the number of requests. Table 6.9 below presents data based on the first 10 entries.[2] 45% of the requests are originated from a few IP addresses.

---

[2] IP addresses and other confidential data were anonymized by LKN before providing the data to RECAP.

Table 6.8 Statistics of the response codes returned by the LKN search engine

HTTP status	Description	Number of entries
200	OK	1,692,070
302	Moved temporarily	175,577
301	Moved permanently	111,109
503	Service unavailable	82,905
404	Not found	45,439
304	Not modified	27,782
206	Partial content	8024
499	Client closed request	2497
403	Forbidden	1190
400	Bad request	699
500	Internal server error	117
405	Method not allowed	100
502	Bad gateway	72
504	Gateway timeout	19

Table 6.9 The top 10 IP addresses directing the largest number of requests to the LKN search engine

Anonymous IP	Count
IP_1	184,381
IP_2	124,093
IP_3a	90,711
IP_3b	85,367
IP_3c	78,317
IP_3d	70,576
IP_4	69,950
IP_5	55,806
IP_6a	44,459
IP_6b	29,771

These IPs correspond to web-spider bots from large search companies, e.g. Google and Yandex. As this provides visibility for LKN in search engine results, no remediation action was taken.

Response Size and Response Time

Figure 6.17 shows a histogram of the response volume of the LKN search engine. Figure 6.17(b) presents a histogram of the (total) time that the search engine requires to provide an answer to user queries. The

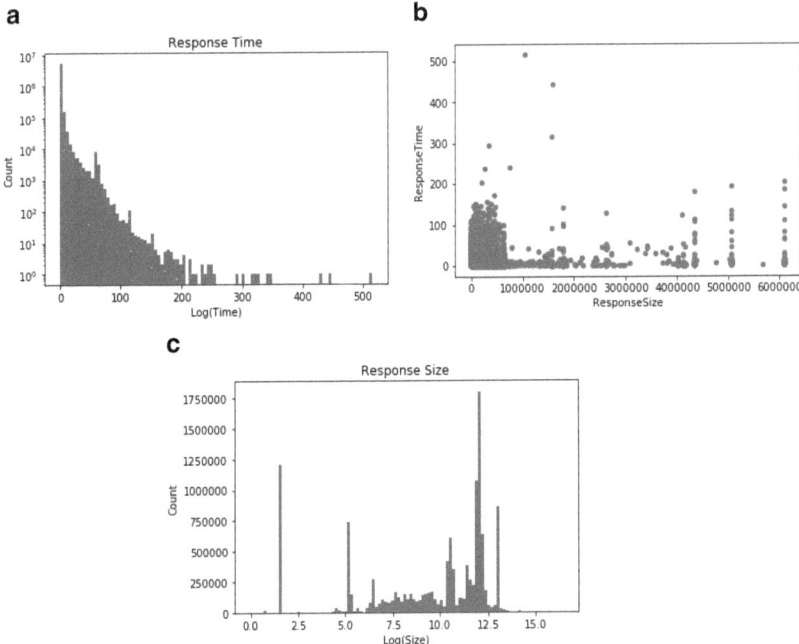

Fig. 6.17 (a), (b), and (c): Histograms of the distribution of the responses, response time, and scatter plot of the response size and time for the LKN search engine

correlation between the response size and the response time was also studied and is shown in Fig. 6.17(c). The Pearson correlation coefficient is 0.18. If only 200 (OK) requests are considered, it is 0.20. Both 0.18 and 0.20 suggest a positive (although weak) correlation between both values.

Number of Requests

The number of requests has also been characterised through a time series with the number of requests that the server receives aggregated over intervals of 30 minutes. This time series is the target workload that is analysed and modelled as part of the RECAP methodology. Peaks in this period can be explained by increased media attention during the period (Fig. 6.18).

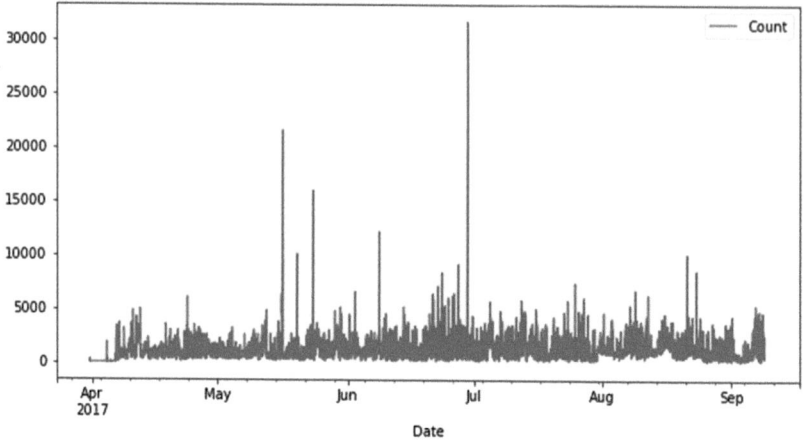

Fig. 6.18 Time series of the LKN's search engine workload (data aggregated over windows of 30 minutes)

6.5.3.2 Workload Predictor Model

From experience, the time to deploy a new data node in Azure clusters is about 30 minutes. The original dataset was aggregated in periods of 30 minutes, and a new feature, number of requests, was derived. Predicting the number of servers required for the next period of 30 minutes (workload prediction) to deal with the expected user requests is the goal of the model.

As a preliminary step before fitting a model to predict future workloads, the stationarity of the time series was examined. A visual inspection of a moving average and a moving standard deviation, together with a decomposition of the series in trends + cycles + noise, suggests a stationary time series. This intuition has been confirmed with a Dickey-Fuller test. The coefficients of the autoregressive integrated moving average (ARIMA) were estimated using an autocorrelation function (autoregressive part) and a partial autocorrelation function (for the moving average) where the already identified components were removed. The final ARIMA model trained corresponds to an autoregressive part of four periods, and a moving average of six periods, without requiring the integration of the original time series.

Figure 6.19 presents a sample dashboard with the Workload Predictor and the Application Modeller for LKN. The dashboard is composed of

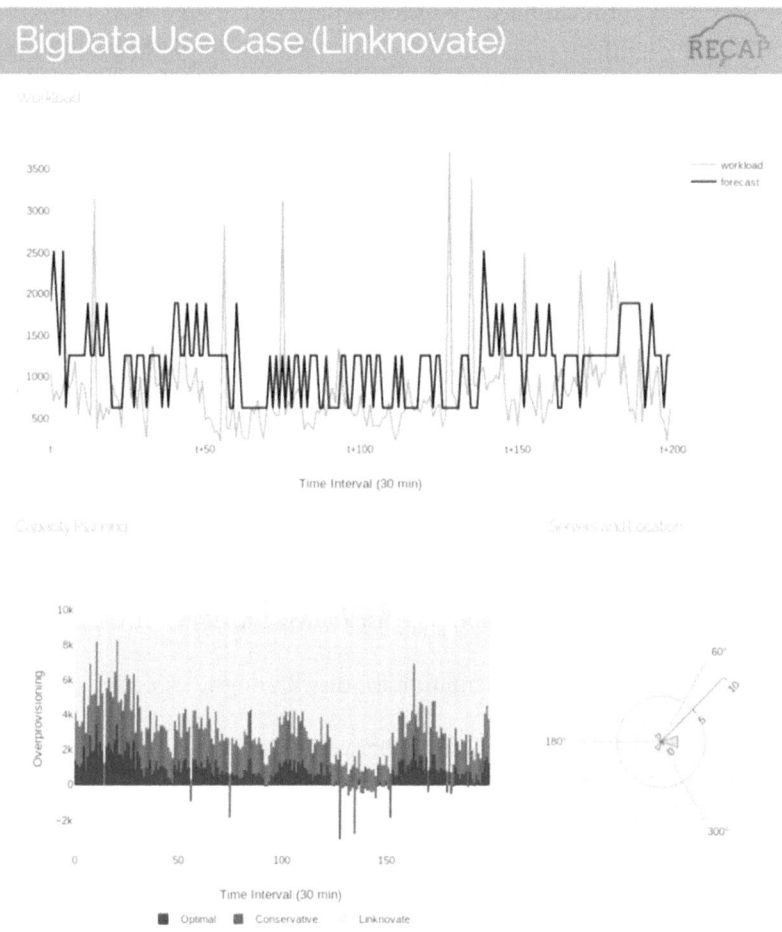

Fig. 6.19 Workload predictor dashboard

three independent but synchronised panes. The "Workload" pane, displayed in light blue, presents the actual workload (number of requests during each 30-minute interval) of the search engine, along with the forecast number of servers required to deal with the predicted workload (in black). The "Capacity Planning" pane presents the overprovisioning of resources deployed to deal with the actual workloads. The current overprovisioning of the servers in production used by LKN is displayed in light blue, while the overprovisioning given the RECAP models is displayed in dark blue together with the underprovisioning in red. The grey display depicts a conservative model that does not underprovision resources. Finally, the "Servers and Location" pane presents the recommended number of servers and their geographical locations, as they are predicted by the application model.

RECAP ran a simulation of the LKN workload based on a historical dataset collected at the production server for a period of one month (August 2017). Based on the collected data, LKN was overprovisioning during that period of time by an average number of 13.8 cores of Azure DS12 v2 i.e. 86.6% of overprovisioning of data processing capacity. Applying the RECAP models would reduce overprovisioning to 3.5 cores, or the equivalent of 60% overprovisioning.

6.5.3.3 RECAP Simulation and Planning Mode

The RECAP Simulation Framework supports a number of features that can help in ES-based system deployment and provisioning decisions. These include:

- Modelling and simulation of a distributed data flow with a hierarchical architecture;
- Custom policy implementation for distributing workload in the hierarchical architecture;
- Synchronous communication between search engine components for data aggregation; and
- Flexible modelling that can be easily adapted to integrate with other CloudSim extensions.

Modelling

Figure 6.20 shows the online virtual layers of the LKN search engine considered in the implementation of the RECAP Simulation Framework. As discussed earlier, the deployed LKN search service stack consists of a web server where the users input their queries and an ES cluster which is responsible for the search and returning the response to the user query. The ES cluster consists of an ES client node and data nodes. The ES node is responsible for: (1) passing and distributing the queries among the data nodes; (2) coordinating and aggregating the search results of different data nodes; (3) and returning the query result to the web server, which in turn returns it to the user. The data nodes are responsible for storing and processing old and fresh data.

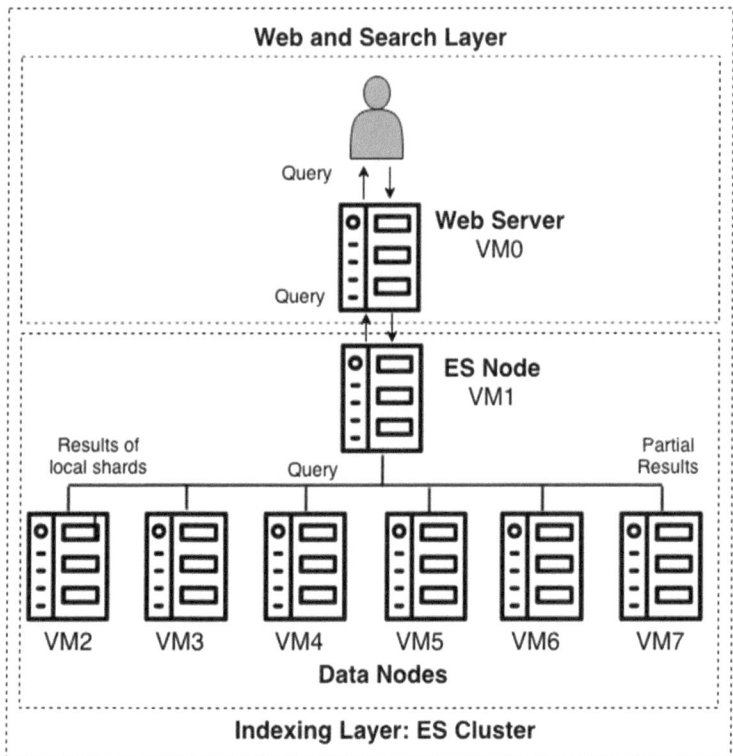

Fig. 6.20 LKN conceptual ElasticSearch (ES) architecture

A simulation model was built to reflect the behaviour of a real ES-based system deployed in a public cloud based on the LKN workload data as a reference. An ES-based search engine was then modelled and simulated using a Discrete Event Simulation (DES) approach. To do so, CloudSim, a widely used open source DES platform, was extended with the simulation model and then compared with KPI traces collected from LKN.

Figure 6.21 illustrates the ES workload flow. Within CloudSim modelling concepts, a cloudlet represents a task submitted to a cloud environment for processing. When a query is launched, a set of cloudlets is generated and executed in sequential manner. The first cloudlet is executed at a web server then the second cloudlet is executed at the ES node. From the ES node, a set of cloudlets (which is less or equal to the number

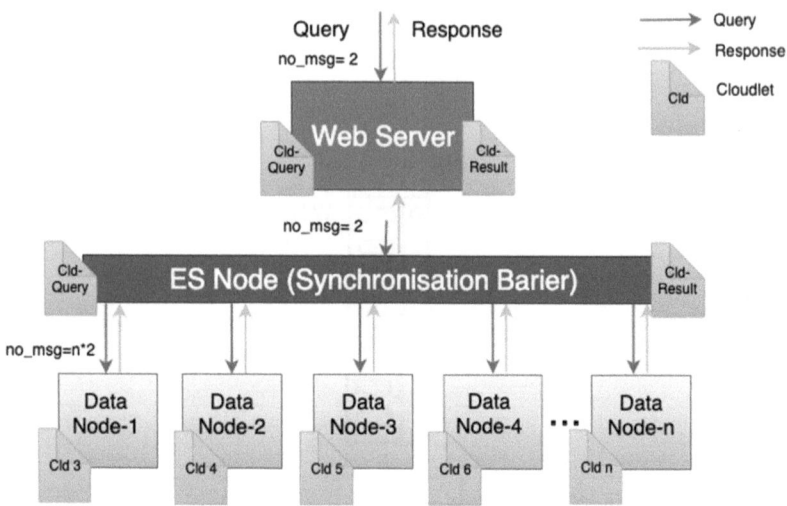

Fig. 6.21   ElasticSearch (ES) workload flow

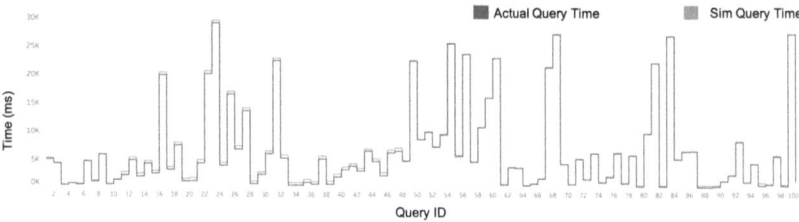

Fig. 6.22   Comparison of actual and simulation query response times

of data nodes) is distributed and executed at data nodes. Next, another cloudlet is executed again at the ES node to merge the partial results coming back from the data nodes. Finally, a last cloudlet goes from the ES node to the web server as a response to the user query.

Results

The simulated response time of a query was compared to its actual time as collected from real system traces. A subset of 100 valid queries was extracted from the data set used. Figure 6.22 compares actual and simulation query response times across the 100 queries. As one can see, the

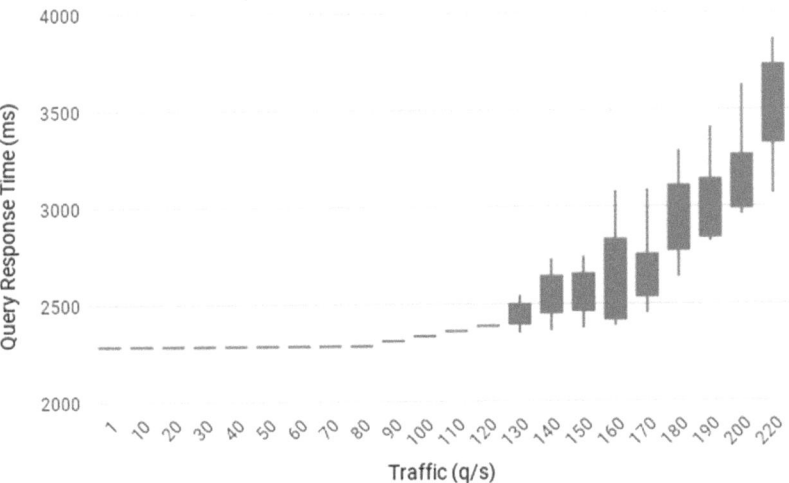

Fig. 6.23 LKN system performance under different traffic scenarios

actual query response times and the simulation query response times are very close and highly positively correlated across all the 100 queries tested.

The performance of the LKN system was analysed by running the simulation with different workloads (query traffic) to see how much traffic the LKN system could handle. Query response time was monitored while varying the number of queries per second (q/s) received by the system. Figure 6.23 is a box plot (min, max, lower quartile, upper quartile) that shows the query response time based on the number of queries per second the system receives.

With query traffic of up to 80 q/s, the query response time for all the queries is the same and it is equal to having one q/s. That means the system is capable of handling 80 q/s with no waiting time. Between 80 q/s and 120 q/s, a slight increase in the response time appears. However, this increase affects all the queries in the same way, i.e. there is no difference in response time between the queries. As we increase the query traffic beyond 120 q/s, a divergence in query response times becomes apparent. Between 130 q/s and 170 q/s, the system manages to execute several queries within a short time by delaying the excess of queries; however, with the increase in query traffic beyond 170 q/s, the system fails to execute even a single query in a short time.

6.5.4   *Results*

The analysis run for this case study has proven valuable in multiple ways. First, LKN was able to gain a better understanding of infrastructure planning, deployment, anomaly detection. Similar, workload prediction models estimate the potential cost savings due to improved resource consumption of 26.6%.

Using the RECAP DES simulator, it was possible to provide an insight for LKN into capacity planning in that it identifies thresholds at which point LKN's QoS starts degrading and additional resources must be provisioned. LKN can now use these data to reduce overprovisioning and at the same time set specific rules for scaling up and down in a cost-effective manner.

All these insights resulted in LKN changing their deployment strategy moving to a more advanced infrastructure configuration using only two data nodes (instead of nine) with NVMe (non-volatile memory express) storage.

References

Ai, Y., M. Peng, and K. Zhang. 2018. Edge Computing Technologies for Internet of Things: A Primer. *Digital Communications and Networks* 4 (2): 77–86.

Ofcom. 2016. Business Connectivity Market Review—Volume I—Review of Competition in the Provision of Leased Lines. https://www.ofcom.org.uk/__data/assets/pdf_file/0015/72303/bcmr-final-statement-volume-one.pdf.

Sauter, T., J. Jasperneite, and L. Lo Bello. 2009. *Towards New Hybrid Networks for Industrial Automation.* 2009 IEEE Conference on Emerging Technologies & Factory Automation, 1–8. IEEE.

INDEX

© The Author(s) 2020   161
T. Lynn et al. (eds.), *Managing Distributed Cloud Applications
and Infrastructure*, Palgrave Studies in Digital Business & Enabling
Technologies, https://doi.org/10.1007/978-3-030-39863-7